Seafood Authenticity and Traceability

Seafood Authenticity and Traceability

A DNA-based Perspective

Edited by

Amanda M. Naaum

Robert H. Hanner

AMSTERDAM • BOSTON • HEIDELBERG • LONDON
NEW YORK • OXFORD • PARIS • SAN DIEGO
SAN FRANCISCO • SINGAPORE • SYDNEY • TOKYO

Academic Press is an imprint of Elsevier

Section II
DNA-Based Analysis for Seafood Authenticity and Traceability

List of Contributors

LeeAnn Applewhite Applied Food Technologies, Inc., Alachua, FL, United States

Christopher Drake Carolin Oceana, Washington, DC, United States

Donna Cawthorn University of Stellenbosch, Stellenbosch, South Africa

Robert H. Hanner University of Guelph, Guelph, ON, Canada

Rosalee S. Hellberg Chapman University, Orange, CA, United States

Johann Hofherr European Commission, Joint Research Centre (JRC)

Patrick Larkin Applied Food Technologies, Inc., Alachua, FL, United States

Stefano Mariani University of Salford, Salford, United Kingdom

Jann Martinsohn European Commission, Joint Research Centre (JRC)

Dana D. Miller University of British Columbia, Vancouver, BC, Canada

Amanda M. Naaum University of Guelph, Guelph, ON, Canada

Einar Eg Nielsen Technical University of Denmark, Silkeborg, Denmark

Sophia J. Pollack Chapman University, Orange, CA, United States

Barbara Rasco Washington State University of Idaho, Pullman, WA, United States

Laurenne Schiller Vancouver Aquarium, Vancouver, BC, Canada

Mahmood S. Shivji Guy Harvey Research Institute/Save Our Seas Shark Research Centre, Dania Beach, FL, United States

U. Rashid Sumaila University of British Columbia, Vancouver, BC, Canada

Eric Enno Tamm This Fish/EcoTrust Canada, Vancouver, BC, Canada

Kimberly Warner Oceana, Washington, DC, United States

Foreword

Humans have relied on the ocean for food for millennia. Today, about 40% of Earth's population lives within 100 km of the coast and over a billion people depend on seafood for their primary source of protein. While in North America the seafood is largely regarded as a luxurious healthy dining option, for millions of people living in developing countries it is an essential dietary requirement. In places such as coastal Africa and the Pacific Islands, fishing also represents a core component of many social systems and cultural activities. In these regions, sustainable seafood is a matter of survival, and the choices we as consumers make ultimately have the potential to affect the health of both the ocean and the communities that depend on it.

Over 7 billion people call Earth home—a threefold increase from only six decades ago. Along with this increase in population has come the emergence of trade networks that connect resources and people around the world, and the benefits and problems associated therein. Looking back to the turn of the 20th century, fishing fleets were largely restricted to coastal regions, and their catch was mainly small, fast-reproducing forage fish, such as sardines and herring. Now, with coastal fish populations collapsing or becoming heavily depleted, and technological advancements allowing fisheries to switch target species and move farther and deeper offshore, consumers of the developed world can eat nearly any seafood their heart desires, regardless of which ocean it is from. Flash-freeze capabilities enable fishers to transport their catch around the world without it spoiling: a tuna caught in Kiribati on a Wednesday can reach a dinner table in Tokyo by Thursday. However, this interconnectivity has dramatically impacted the underlying ecological relationships within the marine environment and, unless management improves, has the potential to affect the resilience of fish species previously unaffected. The global demand for seafood and the associated extent of the world's fishing fleets have increased beyond our capabilities to sustainably manage many of the world's fish stocks. Yet, instead of looking to cease fishing pressure and rebuild depleted fisheries, we have expanded our horizons and palates, allowing a greater diversity of seafood to grace our plates.

Fish is the world's last wild protein hunted at a global scale, and seafood is the most highly traded food commodity in the world. But unlike almost all other food products, such as beef, coffee, and vegetables, identifying fish to the species level can be challenging, if not impossible. There are scores of species—including cod, pollock, haddock, and hake—that are collectively marketed as "whitefish" and, once cooked, a fillet of halibut from the coast of British

Columbia can resemble a cut of mahimahi from Hawaii, or Patagonian toothfish from the Southern Ocean. The situation is further complicated since many species can also go by different appellations depending on where you live. *Dolphin fish* or *dorado* are other terms for mahimahi, and *Chilean seabass* is often used instead of Patagonian toothfish, since the species' original name was believed to sound unappetizing and foreign. In many cases, the use of vague product names can mask the impacts of overexploitation since different fish can be substituted without consumer knowledge of ecological impact. Yet, while consistent product terminology is one reason why traceability is important, having proper information available to consumers is essential for a host of other reasons as well, ranging from human rights abuses and illegal fishing, to the health and safety implications of particular products, especially some farmed ones.

With the longest coastline in the world and shores that touch three different oceans, Canada should be a global leader in marine conservation. Furthermore, since overfishing is no longer a localized issue, all seafood consumers are responsible for ensuring our oceans remain healthy for future generations. In 2005, when the Vancouver Aquarium Marine Science Center launched Ocean Wise, a sustainable seafood program, the overarching goal of this program was not only to raise awareness about how fishing impacts marine ecosystems, but also to reconnect people with the seafood they consume. We realized that the ever-expanding seafood supply chain had resulted in a disconnect between consumers and producers, but that—when provided with sufficient information and guidance—consumers would want to make environmentally responsible choices. Now, with concerns of sustainability at the forefront of many conversations, it can only be expected that more and more people will demand to know the origin of the fish they eat, whether it is wild captured, farmed, local or imported, and its path to their table.

This book provides an introduction to the problem of seafood mislabeling and illegal unreported and unregulated (IUU) fishing, an overview of regulations for traceability of seafood in developed and developing nations, and the importance of consumer choice in seafood management. The book also introduces and reviews the use of DNA analysis of seafood products, one of the most promising approaches for improving authenticity and traceability in this industry. As the seafood supply chain grows ever complex, the traceability principles and initiatives described in this book will be a necessary addition to attaining long-term sustainability in our seafood consumption. It is my greatest hope that the next generation of world citizens will have the ability to not only appreciate the natural beauty of the ocean, but that they will also take responsibility in making the choices required to protecting it. And, demanding information about where and how their seafood was caught will continue to play an important role in attaining this goal.

Dr. John Nightingale
President and CEO
Vancouver Aquarium Marine Science Center

Section I

The Need for Seafood Authenticity

Chapter 1

Seafood Mislabeling Incidence and Impacts

Amanda M. Naaum[1], Kimberly Warner[3], Stefano Mariani[2], Robert H. Hanner[1], Christopher Drake Carolin[3]

[1]*University of Guelph, Guelph, ON, Canada;* [2]*University of Salford, Salford, United Kingdom;* [3]*Oceana, Washington, DC, United States;*

INTRODUCTION

Seafood consumption is at an all-time high globally. In addition to providing almost a quarter of the global intake of animal protein, seafood is also a huge economy, particularly in the developing world, with about one-tenth of the world's population living off revenue derived from fisheries and aquaculture (FAO, 2014). Thousands of species are involved, and the conservation status of wild-harvested species varies both by species and by region. The pressure to meet the global demand for seafood taken together with the biodiversity being exploited and the complexities of global supply chains make it increasingly important, and difficult, to ensure the authenticity and traceability of products on the market. However, misrepresentation exists at many levels. Fraudulent labels may indicate higher weight than actually contained in the package, false points of origin, missing ingredients, and species substitution, all with risks to consumer health and safety (Spink and Moyer, 2011). Labeling regulations vary across regions, but generally some legislation or accepted convention exists in each country that establishes the accepted market names for each seafood product and species. However, a variety of market names may be used for a given species, and a variety of species may be sold under a given market name, even within a single jurisdiction. This situation is further complicated by the fact that different jurisdictions lack harmonization around the application of market names, in part because they use different languages, a situation which challenges accurate labeling in a global market. A key component of regulation is the accurate identification of the fish being caught, imported, and sold to consumers at retail outlets or restaurants. Once seafood products have been processed and packaged, it can be difficult to accurately identify them to species because the morphological characteristics used to identify them with traditional methods are lacking. New DNA-based tools can circumvent this weakness and enhance the accuracy of seafood labeling by validating the

Seafood Authenticity and Traceability. http://dx.doi.org/10.1016/B978-0-12-801592-6.00001-2

identity of the species in the package with market name under which a product is being sold. This chapter reviews a growing body of literature describing incidences of seafood species mislabeling from market surveys, and details their economic, conservation, and health impacts.

GENERAL TRENDS

Evidence of seafood mislabeling has been confirmed as far back as 1915 in a newspaper article that reported excess catches of sharks were to be sold as swordfish (Anonymous, 1915). This still continues to be an issue today as documented by market surveys, which have been increasing in number over the past eight years, as outlined in a 2014 global review (Golden and Warner, 2014). In general, we have considered market surveys to refer to studies of mislabeling at various points in the supply chain comparing labeled market names to those identified by analysis, and determining if these correspond based on accepted legislation or convention for the product (Fig. 1.1). To date, every study

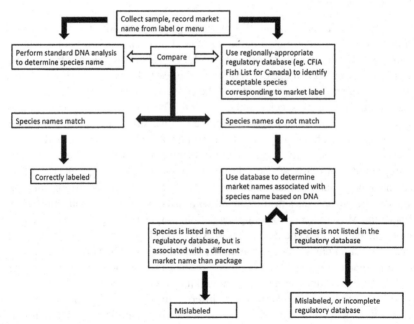

FIGURE 1.1 Process for assessing species authenticity of market samples to identify mislabeling. Samples are collected and identification is made based on DNA analysis. When available, an appropriate regulatory database should be used to ensure that this identification matches the species linked to the market name associated with the original sample. Matches or clear mismatches represent correctly labeled and mislabeled samples, respectively. However, if the species identified using DNA analysis is not listed in the regulatory database, this does not necessarily imply mislabeling. At times the regulatory database may not be fully up to date, and therefore further research into the reason for the missing information is warranted before determining that sample to be mislabeled.

examining market samples for species authenticity has found some level of mis-labeling except for one small study in Tasmania (Lamendin et al., 2015). Levels of mislabeling have ranged from 5% or less (Herrero et al., 2011b; Huxley-Jones et al., 2012; Bernard-Capelle et al., 2014; Rasmussen Hellberg et al., 2011) to 100% of samples (Armani et al., 2012, 2013; Fuller, 2007; Gold et al., 2011), but are generally between 20 and 25%. Some studies collect just a handful of samples (Gold et al., 2011; Burros, 2005), often to demonstrate the effectiveness of new genetic identification techniques, while others analyze over a thousand (Fain et al., 2013; Warner et al., 2013). Despite ample media attention surrounding the issue of seafood fraud, studies conducted in Florida suggest that levels of mislabeling may not have changed very much over time (Warner et al., 2012a) while a long-term study in the USA determined the aver-age level of seafood mislabeling to be 37% (Tennyson et al., 1997). The situa-tion in western Europe may be improving, as rates of mislabeling appear to be declining (Mariani et al., 2015), at least for some species. Most market studies were conducted after 2005, because the increased availability and simplicity of DNA-based testing (Golden and Warner, 2014) coupled with the rapid expansion of available DNA reference sequences made such testing broadly accessible. DNA barcoding is currently the most common method for species identification in market surveys (Griffiths et al., 2014; Rasmussen Hellberg and Morrisey, 2011), but visual identification (e.g., simply comparing package labels to menu items) is still often used to identify fraud in restaurant inspections (Hong, 2014). (Table 1.1) summarizes incidence and impact of mislabeling found in published data from market surveys conducted using DNA barcoding prior to 2015.

DNA analysis, such as DNA barcoding, is particularly useful for processed prod-ucts where morphological features are no longer present. In seafood products, once the distinguishing morphological features present on a whole fish have been removed, identification using traditional methods (e.g., using keys) can be difficult or impos-sible. In very highly processed products, such as canned samples, DNA barcoding does not perform efficiently (Haye et al., 2012), however other methods, such as real-time PCR, targeting a shorter region of DNA, can still be employed to conduct a market survey (Rasmussen Hellberg et al., 2011, Brechon et al., 2013; Espeneira and Vietes, 2012; Hererro et al., 2011a). Some market surveys indicate higher levels of mislabeling in processed products compared to whole fish (Anonymous, 2011; Di Pinto et al., 2013; Espineira et al., 2008a,b; Keskin and Atar, 2012; Pepe et al., 2007). Certainly the lack of morphological features and the modifications inherent to the processing methods make mislabeling easier. However, other surveys cite little to no mislabeling in processed goods (Aguilera-Munoz et al., 2008; Brechon et al. 2013; Huxley-Jones et al., 2012; Bernard-Capelle et al., 2014; Rasmussen Hellberg et al., 2011). In many countries, seafood species labeling regulations are often more lax when seafood is heavily processed, for example, as outlined in the USA Country of Origin Labeling for Fish and Shellfish 7 C.F.R. Part 60 and the new fish and aquaculture labels in the EU (http://ec.europa.eu/fisheries/documentation/publications/eu-new-fish-and-aquaculture-consumer-labels-pocket-guide_en.pdf).

TABLE 1.1 Summary of Results From DNA Barcoding Market Surveys that Specifically State DNA Barcoding Using COI as a Method for Identification

Year	Location	Type of Publication	Number of Samples	Percentage Mislabeled	Impact of Mislabeling	Citations
2011	USA	Media article	183	48%	E, H	Abelson and Daley (2011)
2012	USA	Media article	76	76%	E, C, H	Abelson and Daley (2012)
2011	Ireland	Research group	111	19%	E	Anonymous (2011)
2010	Brazil	Journal article	22	36%	E	Ardura et al. (2010)
2010	Italy	Journal article	40	85%	E	Barbuto et al. (2010)
2011	UK	Media article	21a	19%	C	Boucher (2011)
2011	Brazil	Journal article	63	78%	E	Carvalho et al. (2011)
2012	South Africa	Journal article	248	21%	E, C	Cawthorn et al. (2012)
2013	Iran	Journal article	27	11%	E, H	Changizi et al. (2013)
2012	USA	Journal article	99	20%	E, H	Cline (2012)
2011	USA	Research group	190	18%	E, H	Consumer Reports (2011)
2014	Italy	Journal article	18	17%	E	Cutarelli et al. (2014)
2013	Italy	Journal article	65	31%	E, C	Di Pinto et al. (2013)
2010	Italy	Journal article	69	32%	E, H, C	Filonzi et al. (2010)
2013	UK / Ireland	Journal article	6b	33%	C	Griffiths et al. (2013)
2011	Canada	Journal article	230	39%	E, C	Hanner et al. (2011)

Year	Country	Source	n	%	Impacts	Reference
2012	Chile	Journal article	114	7%	n/a	Haye et al. (2012)
2012	UK	Journal article	212	1%	n/a	Huxley-Jonse et al. (2012)
2012	Turkey	Journal article	50	84%	n/a	Keskin and Atar (2012)
2009	USA	Journal article	68	7%	H	Lowenstein et al. (2010)
2012	UK/Ireland	Journal article	226	19%	E	Miller et al. (2012)
2011	Italy	Journal article	15	27%	E, H, L	Pappalardo et al. (2011)
2009	USA	Media article	1	100%	E, C	Robinson (2009)
2008	USA	Media article	56	25%	E, C	Stoeckle and Strauss (2008)
2011	USA	Research group	88	18%	C, L	Warner (2011)
2012	USA	Research group	142	38%	E, H, C, L	Warner et al. (2012c)
2013	USA	Research group	1215	33%	E, H, C	Warner et al. (2013)
2012	USA	Research group	119	55%	E, C	Warner et al. (2012b)
2012	USA	Research group	96	30	E, H, C	Warner et al. (2012a)
2008	Canada / USA	Journal article	91	25%	E, H, C	Wong and Hanner (2008)
2013	USA	Journal article	7	57%	E, C	Namazie (2014)
2014	Canada	Media article	18	39%	E	Cannon (2014)
2014	Canada	Journal article	294	23%	E, H, C	Naaum and Hanner (2015)
2013	Philippines	Journal article	19	14	E, C	Maralit et al. (2013)

Economic (E), conservation (C), health (H), and lifestyle (L) impacts are listed for studies where they were reported.

The necessity for DNA analysis is partly due to the fact that the opportunity for mislabeling has increased with the expansion of the seafood industry. Seafood is the most highly traded food commodity, and the potential distance seafood travels from the point of capture to consumption has never been higher. Rather than being consumed locally, most seafood is exported from the country where it is caught (FAO, 2014). As products pass through complex supply chains (e.g., involving capture, transshipping, processing, packaging, wholesale, and retail distribution), opportunities for fraud compound, making it difficult to pinpoint when and where mislabeling is occurring. In some media market surveys, which generally focus on samples purchased at grocery stores, fish markets, or restaurants, owners or managers admitted switching restaurant menu labels or labels at fish markets for economic purposes while in other cases, the supplier or other supply chain actors are the perpetrators of species substitutions.

Biological diversity challenges proper labeling of seafood. The number of species being traded is very large and each species may have one or more common names in each region and language where it is consumed. For example, the US Food and Drug Administration (FDA) lists over 1800 species of seafood commercially sold in the USA on their Seafood List (http://www.accessdata.fda.gov/scripts/fdcc/?set=seafoodlist). With so many possibilities available, it is no longer reasonable to assume that a consumer would be able to accurately identify all of the seafood species they are purchasing. Therefore, seafood species mislabeling can be difficult to detect. Adding to the motivation for committing fraud, fines can be very low, in some cases much less than the illicit profits gained (Cohen, 1997). However, public awareness of seafood fraud is increasingly covered in the media, and continued public education may help consumers make more informed decisions.

To help analyze the scope and extent of the seafood fraud problem, Oceana compiled a database of seafood species substitutions reported in market surveys from around the world. This effort complemented the literature review and subsequent mapping of studies identified around the globe (Golden and Warner, 2014 and Google map at http://usa.oceana.org/seafood-fraud/global-reach-seafood-fraud). Database sources include journal articles; research organization-, government-, and journalist-led investigations; and reports of seafood fraud legal cases. Studies identified through early 2015 (138) have been conducted in 30 countries, but the bulk of the studies have been conducted in the USA (41%) and EU (37%). Thus summary statistics reflect a Global North perspective.

From the studies reviewed to date, at least 16,000 individual samples have been analyzed for species authentication with more than 4000 identified as mislabeled. The database contains 313 unique species (and 37 additional taxa identified to the genus or family level) that were found substituted for roughly 100 broad species groups of seafood sold. The exact identity of some (~900) substitute species samples are not known due to the inclusion of journalist, government, and older data that did not identify or report the substitute species. Fraudulent seafood was sold under roughly 200 unique market names, illustrating that people around the globe are exploiting far more species diversity than they realize when consuming seafood. Considering both the relatively high prevalence of global seafood fraud, and the diverse list of aquatic species already being sold and consumed worldwide, the notion that there is an additional, hidden, and biodiverse group of species being consumed unknowingly raises conservation concerns. As this chapter illustrates, among the substitutes were species associated with health and conservation risks in addition to the obvious economic motivations.

The most commonly substituted types of seafood in the database are those labeled as snapper, caviar, grouper, cod, crab, tuna, salmon, and hake, but snapper and cod, by far, are the most studied types of fish in terms of total number of mislabeling studies (33 and 23, respectively, thus far). The most common substitutes appearing across multiple studies are Asian catfish *(Pangasius)*, hake, farmed salmon, shark/rays, tilapia, cod, and escolar/oilfish.

While economically motivated adulteration is likely a contributor to some of the mislabeling identified in market surveys, there is also speculation about other possible causes for fraudulent labels found at fish markets or grocery stores. The first is the lack of explicit regulation governing the labeling of seafood products, which is cited as a complicating factor in some market surveys (Carvalho et al., 2011; Cawthorn et al., 2012; von der Heyden et al., 2010; Lamendin et al., 2015). Even when regulations do exist, market names do not always coincide with only one scientific species name. For example, the FDA lists over 60 species from seven different genera that can be marketed as grouper on their Seafood List. This situation is further complicated by the fact that list of accepted market names may be amended from time to time in order to make species more desirable for consumption (Jacquet and Pauly, 2008). One example of this from the USA is the addition of "Chilean seabass" as an acceptable market name for toothfish (*Dissostichus eleginoides* and *Dissosstichus mawsoni*). Below the national level, local naming conventions for products may also be confusing. For example, under US guidelines, various *Sebastes* spp. can only be marketed as "rockfish" while "red snapper" is reserved for *Lutjanus campechanus*, but within California and Oregon, certain rockfish species are permitted to be marketed as "Pacific red snapper" under California law (Marko et al., 2004; Warner et al., 2012b; Logan et al., 2008; CaL. Code Regs. Tit. 14, §103) or as various snapper names under Oregon law (OR. Rev. Stat. §506.800). In other cases, translations can result in misleading labels. Language barriers and resulting confusion about labeling and traceability requirements may also contribute to noncompliance with regulations (D'Amico et al., 2014). Finally, market names do not always change to reflect the most recent taxonomy, which can affect how product mislabeling is defined. Alaskan (or Walleye) Pollock is one example of this. Due to its highly pelagic streamlined shape, this fish has always been called a "Pollock" and given a separate genus "*Theragra*," but 2006 phylogenetic analysis clearly showed that it is nested within the genus *Gadus*, prompting a renaming (Teletchea et al., 2006). However, the market has not followed biology, and Alaskan Pollock is still largely referred to as this.

Therefore, much of the Alaskan Pollock used to substitute proper "cod" and deemed to constitute mislabeling should be accepted as lawful in those countries that accept "cod" as a name for all *Gadus*. In these types of cases, it can be difficult to differentiate purposeful mislabeling for economic gain from adventitious or unintentional substitutions. However, many cases of substitution appear to involve a species of a lower economic value being substituted for species of a higher value and likely represent cases of economically motivated fraud.

In addition to economic impacts of mislabeled seafood, whether purposeful or not, the social effects of market substitutions can also be detrimental. In some cases, consumers may be searching for products that adhere to certain religious requirements, but product substitution can interfere with this preference. One market survey documented escolar and Asian catfish, which are not kosher, being substituted for albacore and cod, respectively, which are considered kosher (Warner et al., 2012c), while another study suggested that Asian catfish, which cannot be eaten according to Islam, could be employed as a substitute for other species that may be eaten (Changizi et al., 2013). Some products collected in a jellyfish market survey were labeled as bamboo or mustard tube (Armani et al., 2012). The presence of jellyfish in these products may affect consumers who think they are purchasing a vegetarian product. Buying a local product may also be important to consumers, but mislabeling can be used to misrepresent the location where the product was caught from (Pappalardo et al., 2011; Warner, 2011). While these may not represent significant economic impacts, the lifestyle impact is important to consumers, and discovery of these substitutions can negatively affect brand image for companies selling these products, even if the substitution occurred upstream of the retailer in the supply chain. Regardless of where in the supply chain substitution occurs, the impacts are pertinent to consumers, industry, regulators, and the sustainability of the resource.

The following sections summarize some of the most common impacts of seafood mislabeling, including examples of economic, conservation, and human health effects of substitutions from market surveys conducted around the world.

ECONOMIC IMPACT

This is the most commonly cited impact from substitutions found in market surveys because often the substituted fish is of a lower market value than the species listed on the label or menu. The price difference can be quite substantial. One study estimated the cost of substitution of just one species at four million euro (Ardura et al., 2010); another species substitution could cost the US economy seven million dollars a year (Cline, 2012). One example where legal charges were filed resulted from chum salmon (*Oncorhynchus keta*) being sold as king salmon (which refers to Chinook salmon, *Oncorhynchus tshawytscha*), a fish that commands five times the price of chum (Grove, 2012). Red snapper and grouper are two of the products most often mislabeled in market surveys.

The price difference between the genuine fish and their common adulterants can be up to 244% and 214%, respectively, for these high value species (Stiles et al., 2013).

Interestingly, the profits to be made by substitution are highest at wholesale and dockside, but most market surveys are conducted at consumer outlets such as grocery stores and restaurants because this is the only market access point for consumer-driven studies. Additional focus on dockside, import, and wholesale product testing may help to identify the point of substitution in the supply chain, but limited evidence suggests that mislabeling occurs at all steps. For example, seafood substitutions have been found at the wholesale level in the USA and South Africa (FDA, 2014a; Cawthorn et al., 2012) and at the import level in the USA (DOJ, 2002, 2012). At dockside, a large-scale genetic survey of rays landed at EU ports demonstrate that outdated and vague species descriptions and fishery reporting requirements can hide the true identity and conservation status of what is actually being caught and sold (Iglesias et al., 2010). Still, government testing and oversight of seafood supplies is limited. For example, a 2009 audit found that FDA is only able to test about 1% of seafood imports coming into the USA specifically for fraudulent substitutions (GAO, 2009). In addition to more regulatory testing, accessibility of tests or testing services to retailers may help to secure brand image and identify unscrupulous suppliers. This can be accessed through commercial seafood testing services and/or by using commercial kits for directly testing seafood products. Validating species identity can help minimize economic loss from species mislabeling.

CONSERVATION IMPACT

Product mislabeling is one way that illegal, unreported, and unregulated (IUU) products can enter the marketplace and be sold to consumers, particularly in cases where fish are simply labeled as "frozen fillets." A 2009 study provided the first global estimates of illegal and unreported fishing at a level of $10–23 billion annually, representing up to 26 million tonnes of global seafood catches (Agnew et al., 2009). Another 2014 study estimated that more than 20% of wild-caught seafood imported to the USA is a product of illegal and unreported catches (Pramod et al., 2014). Several market surveys have reported the sale of endangered or vulnerable species under other market labels. In one case, critically endangered great white shark was sold under another name (Filonzi et al., 2010). A market survey in Brazil revealed that 24 of 44 samples (55%) of shark purchased in fish markets were actually endangered largetooth sawfish (Melo Malmeira et al., 2013) and a New Zealand study revealed that two of four species of shark found in commercial shark samples were prohibited for sale in the country (Smith and Benson, 2001).

Even when the substituted species is not endangered, incorrectly labeling species content can impact conservation efforts by masking the actual availability of species. This has been cited in market surveys as a potential issue for red snapper (Logan et al., 2008), wild salmon (Burros, 2005), jellyfish (Armani et al., 2013), Italian sardines (Armani et al., 2011), sharks (Barbuto et al., 2010), caviar (Fain et al., 2013), Brazilian catfish (Carvalho et al., 2011), kinglip (Cawthorn et al., 2012), cod (Bréchon et al., 2016), and a freshwater sardine (Maralit et al., 2013). One species of offshore hake, previously not targeted in market surveys because its substitution with silver hake did not represent a significant economic gain, could be harvested at a rate of 1100 metric tons each year based on extrapolation of one study's findings (Garcia-Vasquez et al., 2009). The mislabeling of these species means that (1) the available stock of potentially dwindling species is misrepresented such that it appears abundant to consumers and regulators and (2) species being substituted may be harvested at unsustainable rates, but this is not monitored because they are allowed to enter the market labeled as a different species (Pitcher et al., 2002).

Today, 90% of fisheries are either fully exploited or overexploited (FAO, 2014). To make sustainable choices, consumers rely on available eco-awareness tools (e.g., OceanWise) and labels (e.g., MSC) that suggest preferred seafood options based on conservation status and fishing practices to help them recognize sustainable selections and make informed choices about the products they purchase. However, these tools rely on accurate species identification and labeling in order for consumer choice to exert impacts on the broader market. Unfortunately, even the products these tools recommend to consumers have been found to be mislabeled. Even in usually well-monitored certified supply chains, some instances of mislabeling can occur (e.g., Anderson, 2016). Therefore, product mislabeling also reduces consumer choice and consumer impact on conservation of fish and other seafood.

IMPACT ON HUMAN HEALTH

Seafood caused about 36% of foodborne illness outbreaks in the USA in 2013 (CDC, 2015), and the risk of contracting these illnesses differs among species of fish. Different species may also have varying levels of heavy metals (Lowenstein et al., 2010), toxins (Sapkota et al., 2008), or nutritional value (Weaver et al., 2008; Gribble et al., 2016). Proper labeling helps consumers assess the potential health risks of the food they purchase, but seafood mislabeling hinders this option, exposing consumers to potentially damaging health effects of the unknown species substituted for the products they consume. The health impacts of seafood mislabeling have been documented, and several examples from market surveys illustrate the potential effects of mislabeling, as described below.

In market surveys focused on sushi samples, white tuna is one menu item that is commonly mislabeled. One of the most common substitutes is escolar, *Lepidocybium flavobrunneum*. This species of fish and its close relative, oilfish (*Ruvettus pretiosus*) are known to cause keiorrhea, a wax ester (gempylotoxin) poisoning that occurs because the wax ester in the flesh of these species cannot be properly digested by many individuals (Alexander et al., 2004). Escolar and oilfish (*R. pretiosus*) can also present a risk of histamine poisoning, which can be life threatening (Feldman et al., 2005). These two species have been banned for sale in some countries (e.g., Italy and Japan), but have been found sold under several false market names. Oilfish have caused (gempylotoxin) keiorrhea outbreaks due to mislabeling in Australia and Hong Kong (Gregory, 2002; Ling et al., 2008) while notably, a case of escolar mislabeled as cod received media coverage after causing outbreaks of keiorrhea in consumers in Canada (CBC News, 2007).

Other cases where mislabeling can mask the potential health impacts of product consumption include species that have a higher risk of containing certain toxins. Several cases have been reported in market surveys of shark (Anonymous, 1915; Pappalardo et al., 2011; Warner et al., 2012c) and king mackerel (von der Hyden et al., 2010; Warner et al., 2012a, 2013) being improperly substituted for other species. These species contain higher levels of mercury, and it is recommended that pregnant or nursing women and children avoid these species entirely (FDA, 2014b). Substitution with king mackerel can also pose a health risk due to increased chance of Ciguatera poisoning (Lehane and Lewis, 2000). Pufferfish containing tetrodotoxin has also found its way into markets in the USA labeled as monkfish (Cohen et al., 2009), and in Taiwan in fish powder products (Huang et al., 2014).

The issue of wild-caught versus farmed species is also important from a health perspective. Some farmed species contain different levels and proportions of fatty acids than wild-caught fish for which they would be substituted (Usydus et al., 2011). Therefore substitutions of farmed tilapia, catfish, and salmon for wild fish identified in market surveys can impact consumers who purchase and eat wild-caught fish for the health benefits. In addition, farmed fish may contain higher levels of environmental toxins or other additives used in aquaculture (van Leeuwen et al., 2009). When correctly labeled, testing for these contaminants can be carried out and for this reason, many countries screen imported aquaculture products for banned compounds. However, if mislabeled as wild-caught fish, potentially contaminated aquaculture products may evade import inspection protocols. Hence, mislabeling provides a means for contaminated products of aquaculture to enter the marketplace, circumvent health inspection, and also command a higher price as a wild species, thereby providing a double incentive to cheat.

SEAFOOD MISLABELING TRENDS IN THE UNITED STATES

Compared to global data sources, much of the data for seafood mislabeling in the US come from journalist-led or reported investigations (46%), while peer-reviewed science contributes one-third of the studies. Government, student-based, and research organizations (including Oceana) make up the balance of studies. One large multiyear study of caviar (Fain et al., 2013, 4652 samples) comprises nearly half of total US samples analyzed for seafood substitution in Oceana's database (10172), while 2253 of the database samples were mislabeled. We estimate that 50 broad types of seafood were substituted with at least 167 different species and several more genera/families (much of the journalist reported data are not identified to species).

Of the substitute species identified, over half of the species (53%) and 45% of samples carry a health risk, as determined by the FDA's seafood Hazard Analysis and Critical Control Point (HACCP) guidance (FDA, 2011; HACCP Chapter 3). This means that a large fraction of mislabeled seafood in the USA is likely circumventing species-specific health risk screening measures designed to protect consumers. Eighteen percent of the substitute species carry an extinction risk as determined by the International Union for Conservation of Nature (IUCN). The economic fraud related to farmed for wild substitution not only exposes US consumers to potential health hazards, but may indirectly support human trafficking as some types of farmed and wild seafood used in aquaculture feed are associated with serious human rights abuses, including the use of slavery at some point in the supply chain (Hodal et al., 2014; Mason et al., 2015).

In the US analysis, the types of seafood most commonly mislabeled are sold as snappers, caviar, grouper, tuna, crab, and salmon, but the largest number of studies have been devoted to snapper, grouper, and salmon (Table 1.2). The most common substitutes across multiple studies in the USA are Asian catfish, tilapia, snapper, salmon (farmed), and escolar (Table 1.3). The large majority of snapper fraud involved red snapper specifically, the single most frequently mislabeled seafood species. Our review found 49 different species of fish, many not even sharing a taxonomic family, being sold under the name "red snapper." Of the substitute species in the snapper family, at least 47% (10/21) do not originate in US territorial waters, but range from regions of the world encompassing areas linked to IUU fishing, such as the waters of Indonesia, Madagascar, or Somalia. And while there is certainly no guarantee that these fish were caught illegally, studies since late 2010s have uncovered very strong, species-based, geographic links to IUU fishing. Up to 50% of snappers caught in Indonesian waters, for instance, are estimated to be of IUU origin, and 26% and 43% of the catch in the West and East Indian Oceans (respectively) are estimated as IUU (Pramod et al., 2014; Agnew et al., 2009). As such, seafood fraud and IUU fishing are often intertwined.

AN EMBLEMATIC CASE STUDY OF COD MISLABELING: THE IMPACT OF DNA ANALYSIS

Atlantic cod (*Gadus morhua*) has typically represented a mainstay of human food sources for many centuries (Hutchinson et al., 2015), with a societal role and significance that stretched far beyond the realm of fisheries biology (Kurlansky, 1997). Within the last half decade, cod has also served as an exemplary case

TABLE 1.2 The Types of Seafood Most Commonly Identified as Mislabeled in the USA

Type of Seafood Sold in the USA	N (Studies)	N (Samples)	Study ID#
Snapper	28	445	Wong and Hanner (2008), Jacquet and Pauly (2008), Marko et al. (2004), Hsieh et al. (1995), Vasquez (2009), Robinson (2009), Stoeckle and Strauss (2008), Wolf (2009), Logan et al. (2008), Abelson and Daley (2011), Consumer Reports (2011), Foulke (1993), Garcia and Peele (2012), Abelson and Daley (2012), Namazie (2014), Hsieh (1998), Tennyson et al. (1997), Burros (1992), Grogan (1988, 1989, 1992), Shaikh (2012), Hong (2014), Fielding (2012), Stader (2015), Fox Chicago (2013), and Warner et al. (2013).
Grouper	17	343	Vasquez (2009), Robinson (2009), NOAA (2010), Wolf (2009), Consumer Reports (2011), Abelson and Daley (2012), Nohlgren (2006), Franceschina (2011), Hong (2014), Stader (2015), Reed (2006), DOJ (2012), NOAA (2007), Nohlgren and Tomalin (2006), Nohlgren (2007), Warner et al. (2013), and Strickland and Boyd (2015).
Salmon	12	86	Jacquet and Pauly (2008), Consumer Reports (2006, 2011), Abelson and Daley (2011), Grove (2012), Cline (2012), Burros (2005), Rasmussen Hellberg et al. (2011), Shaikh (2012), Stader (2015), Jackson (2010), and Warner et al. (2013)
Tuna	12	174	Wong and Hanner (2008), Stoeckle and Strauss (2008), Lowenstein et al. (2009), Abelson and Daley (2011, 2012), Consumer Reports (2011), Namazie (2014), Franceshina (2011), Shaikh (2012), Hong (2014), Stader (2015), and Warner et al. (2013)
Caviar	7	327	Jacquet and Pauly (2008), Robinson (2009), Doukakis et al. (2012), Fain et al. (2013), Birstein et al. (1998), Cohen (1997), and DOJ (2002)
Cod	7	57	Jacquet and Pauly (2008), Abelson and Daley (2011, 2012), Burros (1992), Fielding (2012), Stader (2015), and Warner et al. (2013)
Halibut	6	22	Wong and Hanner (2008), Consumer Reports (2011), Namazie (2014), Burros (1992), Fielding (2012), and Warner et al. (2013)
Bass, sea	5	36	Wong and Hanner (2008), Stoeckle and Strauss (2008), Namazie (2014), Stader (2015), and Warner et al. (2013)
Crab	4	161	Wong and Hanner (2008), Hong (2014), Fielding (2012), and Warner et al. (2015)
Bass	4	28	Abelson and Daley (2011, 2012), Stader (2015), and Warner et al. (2013)
Shrimp	1	44	Warner et al. (2014)

TABLE 1.3 The Most Common Types of Seafood Found Substituted in US Seafood Mislabeling Investigations

Type of Substitute Species in the USA	N (Studies)	N (Samples)	Study ID#
Asian catfish	16	62	Wong and Hammer (2008), NOAA (2010, 2007), Wolf (2009), Abelson and Daley (2011), Consumer Reports (2011), Abelson and Daley (2012), Hong (2014), Stader (2015), Reed (2006), DOJ (2012), FoxChicago (2013), Nohlgren (2006, 2007), Warner et al. (2013), and Strickland and Boyd (2015)
Tilapia	14	100	Wong and Hammer (2008), Jacquet and Pauly (2008), Stoeckle and Strauss (2008), Wolf (2009), Logan et al. (2008), Abelson and Daley (2011, 2012), Fuller (2007), Nohlgren and Tomalin (2006), Shaikh (2012), Hong (2014), Fielding (2012), Stader (2015), and Warner et al. (2013)
Snapper	12	136	Wong and Hanner (2008), Marko et al. (2004), Hsieh et al. (1995), Robinson (2009), Stoeckle and Strauss (2008), Abelson and Daley (2011), Consumer Reports (2011), Abelson and Daley (2012), Hsieh (1998), Grogan (1988), Grogan (1992), and Warner et al. (2013)
Salmon	10	82	Jacquet and Pauly (2008), Consumer Reports (2006, 2011), Abelson and Daley (2011), Grove (2012), Cline (2012), Burros (2005), Rasmussen Hellberg et al., (2011), Stader (2015), and Warner et al. (2013)
Escolar	9	136	Lowenstein et al. (2009), Abelson and Daley (2011, 2012), Namazie (2014), Franceshina (2011), Shaikh (2012), Hong (2014), Stader (2015), and Warner et al. (2013)
Caviar/roe	7	325	Jacquet and Pauly (2008), Robinson (2009), Doukakis et al. (2012), Fain et al. (2013), Birstein et al. (1998), Cohen (1997), and DOJ (2002)
Rockfish	7	130	Logan et al. (2008), Foulke (1993), Garcia and Peele (2012), Burros (1992), Fielding (2012), Stader (2015), and Warner et al. (2013)
Bass/sea bass	7	41	Wong and Hanner (2008), Stoeckle and Strauss (2008), Abelson and Daley (2011, 2012), Namazie (2014), Fielding (2012), and Warner et al. (2013)
Seabream	7	29	Wong and Hanner (2008), Robinson (2009), Abelson and Daley (2011), Garcia and Peele (2012), Fuller (2007), Fielding (2012), and Warner et al. (2013)
Cod	5	40	Wong and Hanner (2008), Stoeckle and Strauss (2008), Abelson and Daley (2011, 2012), and Warner et al. (2013)
Crab	2	33	Fielding (2012) and Warner et al. (2013)
Shrimp	1	43	Warner et al. (2014)

study for the impact of DNA technologies on the seafood supply chain. Miller and Mariani (2010) first documented high levels (28%) of cod mislabeling in the city of Dublin, Ireland, which generated substantial media coverage, leading to an unprecedented curb in retail sector operations, as just 2 years later the same retailers sampled in the initial study showed no evidence of species substitution (Mariani et al., 2014). Investigations indicate that Irish retailers are working in a regime of transparency, as no samples from Dublin— and just one from Cork, of 56— proved mislabeled in 2013 (Mariani et al., 2015). At the same time, British retailers went from a 7.8% mislabeling rate in 2010 (Miller et al., 2012) to 2.4% in 2011 (Bréchon et al., 2016), and 5% in 2013 (Mariani et al., 2015), again, indicating a currently rather well-run market sector for this product in the UK. Nevertheless, transnational investigations demonstrate that cod mislabeling rates can vary across countries and depending on product processing (Bréchon et al., 2016; but see also Miller et al., 2012). The biological diversity that "'sustains" cod substitution can also be considerable, with surrogate species ranging from the closely related gadoids (e.g., haddock, saithe, and whiting) to the tropical freshwater catfish *Pangasianodon* and the snailfish family Liparidae (Bréchon et al., 2016). In general, a widely distributed and popularly in-demand product is sustained by a more complex supply network, and hence there is a greater chance for a diversification of the substitute species. Experiences indicate that DNA technology can serve as a catalyst for consumer awareness, updated legislation, and governance efficiency, which are the basis of a transparent and sustainable industry (Mariani et al., 2015). The food service sector (i.e., restaurants and take-away shops) remains much less investigated and likely less effectively regulated (Mariani et al., 2014), posing further pressing challenges to scientists and policy makers.

CONCLUSION

Seafood mislabeling is well documented throughout history, but the breadth and depth of mislabeling is coming into sharper focus, thanks to DNA-based species verification methods. Many market surveys state a lack of, or confusing, regulations as a potential cause of product mislabeling, but new naming standards and regulations can help address this problem (Barendse and Francis, 2015; Presidential Task Force, 2015). For example, in late 2014 the European Union revised seafood-labeling requirements to include both market and scientific names in order to reduce nomenclatural ambiguities. Although the perception exists that consumers do not want this level of information, consumers now have unprecedented access to information via the web and are becoming increasingly interested in and educated about the issues concerning their food supply. Consumer engagement and education is key in mitigating the occurrence of seafood mislabeling, and market surveys are one way for the public to actively participate in learning about seafood labeling. In fact, due to the increased access to DNA-based testing, market surveys can now be carried out relatively easily by interested members of the public. School classes at both the secondary and postsecondary levels have been involved in seafood market surveys (Naaum and Hanner, 2015, Cannon, 2014). The educational benefits of these types of studies have been well characterized (Santschi et al., 2013; Naaum et al., 2014). When conducted in accordance with published standards (Naaum et al., 2015), local studies managed by the public can provide valuable data on the incidence of seafood mislabeling. The potential increase in media coverage of results from market surveys can actually help inform policy changes (Mariani et al., 2014; Presidential Task Force, 2015), positively impacting fisheries regulation and management.

REFERENCES

Abelson, J., Daley, B., October 23, 2011. On the Menu, but Not on Your Plate, the Boston Globe. Retrieved from: http://www.bostonglobe.com/business/2011/10/22/menu-but-not-your-plate/NDbXGXdPR6O37mXRSVPGlL/story.html.

Abelson, J., Daley, B., 2012. Re-testing Mislabeled Fish, Boston Globe. Retrieved from: http://www.bostonglobe.com/Page/Boston/2011-2020/WebGraphics/Business/BostonGlobe.com/2012/11/fish/fish.xml.

Agnew, D.J., Pearce, J., Pramod, G., Peatman, T., Watson, R., Beddington, J.R., Pitcher, T.J., 2009. Estimating the worldwide extent of illegal fishing. PLoS One 4, e4570.

Aguilera-Munoz, F., Valenzuela-Munoz, V., Gallardo-Escárate, C., 2008. Authentication of commercial Chilean mollusks using ribosomal internal transcribed spacer (ITS) as species-specific DNA marker. Gayana 72 (2), 178–187.

Alexander, J., Autrup, H., Bard, D., Carere, A., Guido, L., Cravedi, J., Di Domenico, A., Fanelli, R., Fink-Gremmels, J., Gilbert, J., Grandjean, P., Johansson, N., Oskarsson, A., Renwich, A., Rupricj, J., Schlatter, J., Schoeters, G., Schrenk, D., van Leeuwen, R., Verger, P., 2004. Opinion of the scientific panel on contaminants in the food chain on a request from the commission related to the toxicity of fishery products belonging to the family of Gempylidae. European Food Safety Authority Journal 92, 1–5.

Anderson, L., 2016. From ocean to plate: how DNA testing helps to ensure traceable, sustainable seafood. Marine Stewardship Council. Available: https://www.msc.org/publikacje/z-morza-na-talerz-raport-testy-dna.

Anonymous, September 10, 1915. Shark Meat in Market. New York Times. Retrieved from: http://query.nytimes.com/mem/archive-free/pdf?res=9A03E7DF1239E333A25753C1A96F9C946496D6CF.

Anonymous, 2011. Fish Labelling Survey, Ireland.

Ardura, A., Pola, I.G., Ginuino, I., Gomes, V., Garcia-Vazquez, E., 2010. Application of barcoding to Amazonian commercial fish labelling. Food Research International 43, 1549–1552. http://dx.doi.org/10.1016/j.foodres.2010.03.016.

Armani, A., Castigliego, L., Tinacci, L., Gianfaldoni, D., Guidi, A., 2011. Molecular characterization of icefish, (*Salangidae* family), using direct sequencing of mitochondrial *cytochrome b* gene. Food Control 22, 888–895. http://dx.doi.org/10.1016/j.foodcont.2010.11.020.

Armani, A., D″Amico, P., Castigliego, L., Sheng, G., Gianfaldoni, D., Guidi, A., 2012. Mislabeling of an "unlabelable" seafood sold on the European market: the jellyfish. Food Control 26, 247–251. http://dx.doi.org/10.1016/j.foodcont.2012.01.059.

Armani, A., Tinacci, L., Giusti, A., Castigliego, L., Gianfaldoni, D., Guidi, A., 2013. What is inside the jar? Forensically informative nucleotide sequencing (FINS) of a short mitochondrial

COI gene fragment reveals a high percentage of mislabeling in jellyfish food products. Food Research International. 54 (2), 1383–1393. http://dx.doi.org/10.1016/j.foodres.2013.10.003.

Barbuto, M., Galimberti, A., Ferri, E., Labra, M., Malandra, R., Galli, P., Casiraghi, M., 2010. DNA barcoding reveals fraudulent substitutions in shark seafood products: the Italian case of "palombo" (*Mustelus* spp.). Food Research International 43, 376–381. http://dx.doi.org/10.1016/j.foodres.2009.10.009.

Barendse, J., Francis, J., 2015. Towards a standard nomenclature for seafood species to promote more sustainable seafood trade in South Africa. Marine Policy 53 (C), 180–187. http://dx.doi.org/10.1016/j.marpol.2014.12.007.

Bérnard-Capelle, J., et al., 2014. Fish mislabelling in France: substitution rates and retail types. PeerJ 2, e714.

Birstein, V.J., Doukakis, P., Sorkin, B., DeSalle, R., 1998. Population aggregation analysis of three caviar-producing species of sturgeons and implications for the species identification of black caviar. Conservation Biology 12, 766–775.

Boucher, P., 2011. There's Something Fishy Going On: How Mislabelled Cod Is Slipping Through the Net, the Independent. Retrieved from: http://www.independent.co.uk/life-style/food-and-drink/features/theres-something-fishy-going-on-how-mislabelled-cod-is-slipping-through-the-net-2184097.html.

Brechon, A.L., Coombs, S.H., Sims, D.W., Griffiths, A.M., 2013. Development of a rapid genetic technique for the identification of clupeid larvae in the Western English Channel and investigation of mislabelling in processed fish products. ICES Journal of Marine Science 70 (2), 399–407. http://dx.doi.org/10.1093/icesjms/fss178.

Bréchon, A.L., Hanner, R., Mariani, S., 2016. A systematic analysis across North Atlantic countries unveils subtleties in cod product labelling. Marine Policy. 60, 124–133. http://dx.doi.org/10.1016/j.marpol.2016.04.014.

Burros, M., September 2, 1992. Eating Well: Pollack or Cod? Fish or Foul? F.D.A. Takes a Closer Look. New York Times. http://www.nytimes.com/1992/09/02/garden/eating-well-pollack-or-cod-fish-or-foul-fda-takes-a-closer-look.html?src=pm&pagewanted=1.

Burros, M., 2005. Stores Say Wild Salmon, but Tests Say Farm Bred. New York Times.

Cannon, S., 2014. Some Fish Products Mislabeled, DNA Evidence Suggests. University of Calgary. Retrieved from: http://www.ucalgary.ca/utoday/issue/2014-12-03/some-fish-products-mislabelled-dna-evidence-suggests.

Carvalho, D.C., Neto, D.A.P., Brasil, B.S.A.F., Oliveira, D.A.A., 2011. DNA barcoding unveils a high rate of mislabeling in a commercial freshwater catfish from Brazil. Mitochondrial DNA 22 (S1), 97–105. http://dx.doi.org/10.3109/19401736.2011.588219.

Cawthorn, D.-M., Steinman, H.A., Witthuhn, R.C., 2012. DNA barcoding reveals a high incidence of fish species misrepresentation and substitution on the South African market. Food Research International 46, 30–40. http://dx.doi.org/10.1016/j.foodres.2011.11.011.

CBC News, 2007. Canadians Fall Ill After Eating Mislabelled Oily Fish, CBC News. Retrieved from: http://www.cbc.ca/news/technology/canadians-fall-ill-after-eating-mislabelled-oily-fish-1.649068.

Center for Disease Control and Prevention, 2015. Surveillance for Foodborne Disease Outbreaks, United States, 2013, Annual Report. US Department of Health and Human Services, CDC, Atlanta, Georgia.

Changizi, R., Farahmand, H., Soltani, M., Darvish, F., Elmdoost, A., 2013. Species identification of some fish processing products in Iran by DNA barcoding. Journal of Agricultural Science and Technology 15, 973–980.

Cline, E., 2012. Marketplace substitution of Atlantic salmon for Pacific salmon in Washington State detected by DNA barcoding. Food Research International 45 (1).

Cohen, N.J., Deeds, J.R., Wong, E.S., Hanner, R.H., Yancy, H.F., White, K.D., Gerber, S.I., 2009. Public health response to puffer fish (Tetrodotoxin) poisoning from mislabeled product. Journal of Food Protection 72 (4), 810–817.

Cohen, A., 1997. Sturgeon poaching and black market caviar: a case study. Environmental Biology of Fishes 48, 423–426.

Consumer Reports, 2011. Mystery Fish: The Label Said Red Snapper, the Lab Said Baloney Consumer Reports.

Cutarelli, A., Amoroso, M.G., De Roma, A., Girardi, S., Galiero, G., Guarino, A., Corrado, F., 2014. Italian market fish species identification and commercial frauds revealing by DNA sequencing. Food Control. 37, 46–50. http://dx.doi.org/10.1016/j.foodcont.2013.08.009.

Consumer Reports, 2006. The Salmon Scam: "Wild" Often Isn't Consumer Reports, p. 15.

(DOJ) United States Department of Justice, 2002. Company president pleads guilty to caviar smuggling conspiracy. In: Justice, U. S. D. O. http://www.justice.gov/archive/opa/pr/2002/August/02_enrd_492.htm.

(DOJ) United States Department of Justice, 2012. California seafood corporation sentenced to pay $1 million for false labeling of seafood products. In: United States Department of Justice. http://www.justice.gov/opa/pr/2012/February/12-enrd-171.html.

D'Amico, P., Armani, A., Castigliego, L., Sheng, G., Gianfaldoni, D., Guidi, A., 2014. Seafood traceability issues in Chinese food business activities in the light of the European provisions. Food Control 35, 7–13.

Doukakis, P., Pikitch, E.K., Rothschild, A., DeSalle, R., Amato, G., Kolokotronis, S.-O., 2012. Testing the effectiveness of an international conservation agreement: marketplace forensics and CITES caviar trade regulation. PLoS One 7, e40907.

Espiñeira, M., Vieites, J.M., 2012. Rapid method for controlling the correct labeling of products containing common octopus (*Octopus vulgaris*) and main substitute species (*Eledone cirrhosa* and *Dosidicus gigas*) by fast real-time PCR. Food Chemistry. 135, 2439–2444. http://dx.doi.org/10.1016/j.foodchem.2012.07.056.

Espineira, M., Gonzalez-Lavin, N., Vieites, J.M., Santaclara, F.J., 2008a. Authentication of anglerfish species (*Lophius* spp.) by means of polymerase chain reaction-restriction fragment length polymorphism (PCR-RFLP) and forensically informative nucleotide sequencing (fins) methodologies. Journal of Agricultural and Food Chemistry 56, 10594–10599. http://dx.doi.org/10.1021/jf801728q.

Espineira, M., Gonzalez-Lavin, N., Vieites, J.M., Santaclara, F.J., 2008b. Development of a method for the genetic identification of flatfish species on the basis of mitochondrial DNA sequences. Journal of Agricultural and Food Chemistry 56, 8954–8961. http://dx.doi.org/10.1021/jf800570r.

Fain, S.R., Straughan, D.J., Hamlin, B.C., Hoesch, R.M., LeMay, J.P., 2013. Forensic genetic identification of sturgeon caviars traveling in world trade. Conservation Genetics 14 (4), 855–874. http://dx.doi.org/10.1007/s10592-013-0481-z.

FAO, 2014. The State of World Fisheries and Aquaculture. Retrieved from: http://www.fao.org/resources/infographics/infographics-details/en/c/231544/.

Feldman, K.A., Werner, S.B., Cronan, S., Hernandez, M., Horvath, A.R., Lea, C.S., Au, A.M., Vugia, D.J., 2005. A large outbreak of scombroid fish poisoning associated with eating escolar fish (*Lepidocybium flavobrunnem*). Epidemiology and Infection 133 (1), 29–33.

Fielding, J., November 1, 2012. Mislabeled seafood sold in restaurants and grocery stores. In: Los Angeles Department of Public Health, Los Angeles, California. http://file.lacounty.gov/bc/q2_2012/cms1_178178.pdf.

Filonzi, L., Chiesa, S., Vaghi, M., Marzano, F.N., 2010. Molecular barcoding reveals mislabelling of commercial fish products in Italy. Food Research International 43, 1383–1388. http://dx.doi.org/10.1016/j.foodres.2010.04.016.

Foulke, J., 1993. Is Something Fishy Going On? Intentional Mislabeling of Fish, FDA Consumer. U.S. Government Printing Office.

Fox Chicago, August 21, 2013. Addison Seafood Distributor Fined for Mislabeling Fish, Shrimp. Fox Chicago, Chicago. http://www.myfoxchicago.com/story/23075686/addison-seafood-distributor-fined-for-mislabeling-fish-shrimp#ixzz2kdn9MZ8g.

Franceschina, P., June 16, 2011. Fish Fraud Means What's on Your Plate May Be an Impostor. Sun Sentinel, Miami. http://articles.sun-sentinel.com/2011-06-16/news/fl-fake-fish-supplies-20110528_1_mislabeled-fish-fish-fraud-seafood-fraud.

Fuller, J.R., May 10, 2007. Hook, Line & Stinker; the Menus Said Snapper. But it Wasn't!. Chicago Sun Times, Chicago.

GAO United States Governament Accountability Office, 2009. Seafood Fraud: FDA Program Changes and Better Collaboration Among Key Federal Agencies Could Improve Detection and Prevention. United States Government Accountability Office, Washington, DC, pp. 09–258.

Garcia, A., Peele, R., October 26, 2012. Get Garcia, Get Results: Mislabeled Fish Uncovered. NBC, Los Angeles. http://www.nbclosangeles.com/investigations/series/get-garcia/Get-Ana-Garcia-Get-Results-Fish-Investigation-Labeling-Food-Health-Hidden-Camera-175915241.html.

Garcia-Vazquez, E., Horreo, J.L., Campo, D., Machado-Schiaffino, G., Bista, I., Triantafyllidis, A., Juanes, F., 2009. Mislabeling of two commercial North American hake species suggests underreported exploitation of offshore hake. Transactions of the American Fisheries Society 138 (4), 790–796. http://dx.doi.org/10.1577/T08-169.1.

Gold, J.R., Voelker, G., Renshaw, M.A., 2011. Phylogenetic relationships of tropical western Atlantic snappers in subfamily *Lutjaninae* (*Lutjanidae*: Perciformes) inferred from mitochondrial DNA sequences. Biological Journal of the Linnean Society 102 (4), 915–929. http://dx.doi.org/10.1111/j.1095-8312.2011.01621.x.

Golden, R.E., Warner, K., 2014. The Global Reach of Seafood Fraud: A Current Review of the Literature. From: https://s3.amazonaws.com/s3.oceana.org/images/Seafood_Fraud_Map_White_paper_new.pdf.

Gregory, J., 2002. Outbreaks of diarrhoea associated with butterfish in Victoria. CDI 26.

Gribble, M.O., Karimi, R., Feingold, B.J., Nyland, J.F., O'Hara, T.M., Gladyshev, M.I., Chen, C.Y., 2016. Mercury, selenium and fish oils in marine food webs and implications for human health. Journal of the Marine Biological Association of the United Kingdom 96, 43–59.

Griffiths, A.M., Fox, J., Greenfield, A., Miller, D.D., Egan, A., Mariani, S., 2013. DNA barcoding unveils skate (Chondrichthyes: Rajidae) species diversity in "ray" products sold across Ireland and the UK. PeerJ 1 (129), 1–12. http://dx.doi.org/10.7717/peerj.129.

Griffiths, A.M., Sotelo, C.G., Mendes, R., Pérez-Martín, R.I., Schröder, U., Shorten, M., Silva, H.A., Verrez-Bagnis, V., Mariani, S., 2014. Food Control 45, 95–100.

Grogan, J., August 22, 1988. Seafood Industry Scrambles to Recover - Store Owners Take Steps to Ease Customer Fears. Sun Sentinel. http://articles.sun-sentinel.com/1988-08-22/news/8802180412_1_seafood-industry-triple-m-seafood-snapper.

Grogan, J., February 12, 1989. State Tests Find Fake Red Snapper, Fishy Labels. Sun Sentinel. http://articles.sun-sentinel.com/1989-02-12/news/8901080812_1_snapper-mislabeled-samples.

Grogan, J., June 18, 1992. Seafood Labeling Plenty Fishy, State Study Says. Sun Sentinel, Fort Lauderdale, FL. http://articles.sun-sentinel.com/1992-06-18/news/9202150226_1_red-snapper-mislabeled-seafood.

Grove, C., 2012. Fish Sold as King Salmon Turn Out to Be Chums; Seller Faces Felony Conviction. Anchorage Daily News.

Hanner, R., Becker, S., Ivanova, N.V., Steinke, D., 2011. FISH-BOL and seafood identification: geographically dispersed case studies reveal systemic market substitution across Canada. Mitochondrial DNA 22 (Suppl. 1), 106–122. http://dx.doi.org/10.3109/19401736.2011.588217.

Haye, P.A., Segovia, N.I., Vera, R., Gallardoa, M.d.l.Á., Gallardo-Escárate, C., 2012. Authentication of commercialized crab-meat in Chile using DNA Barcoding. Food Control. 25 (1), 239–244. http://dx.doi.org/10.1016/j.foodcont.2011.10.034.

Herrero, B., Lago, F.C., Vieites, J.M., Espineira, M., 2011a. Authentication of swordfish (*Xiphias gladius*) by RT–PCR and FINS methodologies. European Food Research and Technology 233 (2), 195–202. http://dx.doi.org/10.1007/s00217-011-1502-0.

Herrero, B., Vieites, J.M., Espineira, M., 2011b. Authentication of Atlantic salmon (*Salmo salar*) using real-time PCR. Food Chemistry 127 (3), 1268–1272. http://dx.doi.org/10.1016/j.foodchem.2011.01.070.

Hodal, K., Kelly, C., Lawrence, F., June 10, 2014. Asian Slave Labour Producing Prawns for Supermarkets in US, UK the Guaridian, UK. http://www.theguardian.com/global-development/2014/jun/10/supermarket-prawns-thailand-produced-slave-labour.

Hong, C., April 29, 2014. At Some Restaurants, the Fish on the Plate Isn't Always What the Customer Orders. Florida Times Union, Duval, Florida. http://members.jacksonville.com/news/florida/2014-04-29/story/fish-plate-isnt-always-what-customer-orders.

Hsieh, Y.-H.P., 1998. Species substitution of restaurant fish entrees. Journal of Food Quality. 21 (1), 1–11. http://dx.doi.org/10.1111/j.1745-4557.1998.tb00499.x.

Hsieh, Y.-H.P., Woodward, B.B., Blanco, A.W., 1995. Species substitution of retail snapper fillets. Journal of Food Quality 18, 131–140.

Huang, Y.-R., Yin, M.-C., Hsieh, Y.-L., Yeh, Y.-H., Yang, Y.-C., Chung, Y.-L., Hsieh, C.-H.E., 2014. Authentication of consumer fraud in Taiwanese fish products by molecular trace evidence and forensically informative nucleotide sequencing. Food Research International 55, 294–302. http://dx.doi.org/10.1016/j.foodres.2013.11.027.

Hutchinson, W.F., Culling, M., Orton, D.C., Hänfling, B., Handley, L.L., Hamilton-Dyer, S., O'Connell, T.C., Richards, M.P., Barret, J.H., 2015. The globalization of naval provisioning: ancient DNA and stable isotope analyses of stored cod from the wreck of the Mary Rose, AD 1545. Royal Society Open Science. http://dx.doi.org/10.1098/rsos.150199.

Huxley-Jones, E., Shaw, J.L.A., Fletcher, C., Parnell, J., Watts, P.C., 2012. Use of DNA barcoding to reveal species composition of convenience seafood. Conservation Biology 26 (2), 367–371.

von der Heyden, S., Barendse, J., Seebregts, A.J., Matthee, C.A., 2010. Misleading the masses: detection of mislabelled and substituted frozen fish products in South Africa. ICES Journal of Marine Science 67, 176–185.

Iglesias, S.P., Toulhoat, L., Sellos, D.Y., 2010. Taxonomic confusion and market mislabelling of threatened skates: important consequences for their conservation status. Aquatic Conservation: Marine and Freshwater Ecosystems 20, 319–333.

Jackson, D., May 26, 2010. Gastonia Company Admits to Mislabeling Fish Sold to Grocery Chain, Gaston Gazette Gaston County, North Carolina. http://www.gastongazette.com/articles/grocery-47597-food-sent.html#ixzz24TdZGxQs.

Jacquet, J.L., Pauly, D., 2008. Trade secrets: renaming and mislabeling of seafood. Marine Policy 32, 309–318. http://dx.doi.org/10.1016/j.marpol.2007.06.007.

Keskin, E., Atar, H.H., 2012. Molecular identification of fish species from surimi-based products labeled as Alaskan pollock. Journal of Applied Ichthyology 28, 811–814. http://dx.doi.org/10.1111/j.1439-0426.2012.02031.x.

Kurlansky, M., 1997. Cod: A Biography of the Fish that Changed the World. Walker Publishing Company, New York.

Lamendin, R., Miller, K., Ward, R.D., 2015. Labelling accuracy in Tasmanian seafood: an investigation using DNA barcoding. Food Control 47, 436–443.

Lehane, L., Lewis, R.J., 2000. Ciguatera: recent advances but the risk remains. International Journal of Food Microbiology 61, 91–125.

Ling, K.H., Cheung, C.W., Cheng, S.W., Cheng, L., Li, S.-L., Nichols, P.D., Ward, R.D., Graham, A., But, P.P.-H., 2008. Rapid detection of oilfish and escolar in fish steaks: a tool to prevent keriorrhea episodes. Food Chemistry 110, 538–546.

Logan, C.A., Alter, S.E., Haupht, A.J., Tomalty, K., Palumbi, S.R., 2008. An impediment to consumer choice: overfished species are sold as Pacific red snapper. Biological Conservation 141 (6), 1591–1599. http://dx.doi.org/10.1016/j.biocon.2008.04.007.

Lowenstein, J.H., Amato, G., Kolokotronis, S.O., 2009. The real maccoyii: identifying tuna sushi with DNA barcodes–contrasting characteristic attributes and genetic distances. PLoS One 4, e7866.

Lowenstein, J.H., Burger, J., Jeitner, C.W., Amato, G., Kolokotronis, S., Gochfeld, M., 2010. DNA barcodes reveal species-specific mercury levels in tuna sushi that pose a health risk to consumers. Biology Letters 6, 692–695.

van Leeuwen, S.P.J., van Valzen, M.J.M., Swart, C.P., van der Veen, I., Traag, W.A., de Boer, J., 2009. Halogenated contaminants in farmed salmon, trout, tilapia, pangasius, and shrimp. Environmental Science and Technology 43 (11), 4009–4015.

Maralit, B.A., Aguila, R.D., Ventoleroa, M.F.H., Perez, S.K.L., Willette, D.A., Santos, M.D., 2013. Detection of mislabeled commercial fishery by-products in the Philippines using DNA barcodes and its implications to food traceability and safety. Food Control. 33, 119–125. http://dx.doi.org/10.1016/j.foodcont.2013.02.018.

Mariani, S., Ellis, J., O'Reilly, A., Brechon, A.L., Sacchi, C., Miller, D.D., 2014. Mass media influence and the regulation of illegal practices in the seafood market. Conservation Letters. http://dx.doi.org/10.1111/conl.12085.

Mariani, S., Griffiths, A.M., Velasco, A., Kappel, K., Jérôme, M., Perez-Martin, R.I., Schröder, U., Verrez-Bagnis, V., Silva, H., Vandamme, S.G., Boufana, B., Mendes, R., Shorten, M., Smith, C., Hankard, E., Hook, S.A., Weymer, A.S., Gunning, D., Sotelo, C.G., 2015. Low mislabeling rates indicate marked improvements in European seafood market operations. Frontiers in Ecology and the Environment 13, 536–540. http://dx.doi.org/10.1890/150119.

Marko, P.B., Lee, S.C., Rice, A.M., Gramling, J.M., Fitzhenry, T.M., McAlister, J.S., Moran, A.L., 2004. Fisheries: mislabelling of a depleted reef fish. Nature 430 (6997), 309–310. http://dx.doi.org/10.1038/430309b.

Mason, M., Mendoza, M., McDowell, R., March 25, 2015. AP Investigation: Are Slaves Catching the Fish You Buy? March 15, 2015, Star Tribune. Associated Press, Minneapolis, MN. http://www.startribune.com/ap-investigation-is-the-fish-you-buy-caught-by-slaves/297484221/.

Melo Palmeira, C.A., Silva Rodrigues-Filho, L.F.d., Luna Sales, J.B.d., Vallinoto, M., Schneider, H., Sampaio, I., 2013. Commercialization of a critically endangered species (largetooth sawfish, Pristis perotteti) in fish markets of northern Brazil: authenticity by DNA analysis. Food Control. 34, 249–252. http://dx.doi.org/10.1016/j.foodcont.2013.04.017.

Miller, D.D., Mairani, S., 2010. Smoke, mirrors and mislabeled cod: poor transparency in the European seafood industry. Frontiers in Ecology and the Environment 8, 517–521.

Miller, D., Jessel, A., Mariani, S., 2012. Seafood mislabelling: comparisons of two western European case studies assist in defining influencing factors, mechanisms and motives. Fish and Fisheries. http://dx.doi.org/10.1111/j.1467-2979.2011.00426.x.

Naaum, A.M., Hanner, R., 2015. Community engagement in seafood identification using DNA barcoding reveals market substitution in Canadian seafood. DNA Barcodes 3, 74–79.

Naaum, A.M., Frewin, A., Hanner, R., 2014. DNA barcoding as an educational tool: case studies in insect biodiversity and seafood identification. Teaching and Learning Innovations Journal 16 journal.lib.uoguelph.ca/index.php/tli/article/view/2790.

Naaum, A.M., St Jaques, J., Warner, K., Santshi, L., Imondi, R., Hanner, R., 2015. Standards for conducting a DNA barcoding market survey: minimum information and best practices. DNA Barcodes 3, 80–84.

Namazie, Y., 2014. Something's Fishy: Why Your Sushi May Not Be Authentic, Oracle. Retrieved from: http://oracle.myarcher.org/?p=7716.

National Oceanic and Atmospheric Administration, 2007. Seafood importer and associated corporations receive imprisonment and fines. Retrieved from: http://www.publicaffairs.noaa.gov/releases2007/jan07/noaa07-r101.html.

NOAA Office of Law Enforcement, 2010. Three individuals indicted for false labeling, smuggling, and misbranding of seafood products. In: United States National Oceanic and Atmospheric Administration, N. O. A. A. http://www.nmfs.noaa.gov/ole/news/news_sed_012810.htm.

Nohlgren, S., Tomalin, T., August 6, 2006. You Order Grouper; What Do You Get? Tampa Bay Times. http://www.sptimes.com/2006/08/06/Tampabay/You_order_grouper_wha.shtml.

Nohlgren, S., December 6, 2006. How to Prove It's Grouper? Tampa Bay Times. http://www.sptimes.com/2006/12/06/State/How_to_prove_it_s_gro.shtml.

Nohlgren, S., January 29, 2007. State finds more grouper impostors. Tampa Bay Times. Retrieved from: http://www.sptimes.com/2007/01/29/.

Di Pinto, A., Di Pinto, P., Terio, V., Bozzo, G., Bonerba, E., Ceci, E., Tantillo, G., 2013. DNA barcoding for detecting market substitution in salted cod fillets and battered cod chunks. Food Chemistry. 141, 1757–1762. http://dx.doi.org/10.1016/j.foodchem.2013.05.093.

Pappalardo, A.M., Guarino, F., Reina, S., Messina, A., De Pinto, V., 2011. Geographically widespread swordfish barcode stock identification: a case study of its application. PLoS One 6 (10), e25516. http://dx.doi.org/10.1371/journal.pone.0025516.

Pepe, T., Trotta, M., di Marco, I., Anastasio, A., Bautista, J.M., Cortesi, M.L., 2007. Fish species identification in surimi-based products. Journal of Agricultural and Food Chemistry 55, 3681–3685. http://dx.doi.org/10.1021/jf063321o.

Pitcher, T.J., Watson, R., Forrest, R., Valtysson, H., Guenette, S., 2002. Estimating illegal and unreported catches from marine ecosystems: a basis for change. Fish and Fisheries 3, 317–339.

Pramod, G., Nakamura, K., Pitcher, T.J., Delagranc, L., 2014. Estimates of illegal and unreported fish in seafood imports to the USA. Marine Policy 48, 102–113. http://dx.doi.org/10.1016/j.marpol.2014.03.019.

Presidential Task Force on Combating IUU Fishing and Seafood Fraud, 2015. In: Action Plan for Implementing the Task Force Recommendations in: NOAA Fisheries, International Affairs. http://www.nmfs.noaa.gov/ia/iuu/noaa_taskforce_report_final.pdf.

Rasmussen Hellberg, R.S., Morrisey, M.T., 2011. Advances in DNA-based techniques for the detection of seafood species substitution on the commercial market. Journal of Laboratory Automation 16, 308–321.

Rasmussen Hellberg, R.S., Naaum, A.M., Handy, S.M., Hanner, R.H., Deeds, J.R., Yancy, H.F., Morrisey, M.T., 2011. Interlaboratory evaluation of a real-time multiplex polymerase chain reaction method for identification of salmon and trout species in commercial products. Journal of Agricultural and Food Chemistry 59, 876–884. http://dx.doi.org/10.1021/jf103241y.

Reed, M., November 21, 2006. Florida Restaurants Fight Off Fake Grouper. USA Today. http://usatoday30.usatoday.com/news/nation/2006-11-21-florida-fake-grouper_x.htm.

Robinson, J., 2009. Through DNA Testing, Two Students Learn What's What in Their Neighborhood. New York Times. Retrieved from: http://query.nytimes.com/gst/fullpage.html?res=9A03E0DE1130F93BA15751C1A96F9C8B63.

Santschi, L., Hanner, R.H., Ratnasingham, S., Riconscente, M., Imondi, R., 2013. Barcoding Life's Matrix: translating biodiversity genomics into high school settings to enhance life science education. PLoS Biology 11, e1001471. http://dx.doi.org/10.1371/journal.pbio.1001471.

Sapkota, A., Sapkota, A.R., Kucharski, M., Burke, J., McKenzie, S., Walker, P., Lawrence, R., 2008. Aquaculture practices and potential human health risks: Current knowledge and future priorities. Environment International 34, 1215–1226.

Shaikh, S., 2012. Educational posters: mislabeling in the sushi industry. Vitro Cellular & Developmental Biology - Animal, 48, pp. S53–S54.

Smith, P.J., Benson, P.G., 2001. Biochemical identification of shark fins and fillets from the coastal fisheries in New Zealand. Fishery Bulletin 99 (2), 351–355.

Spink, J., Moyer, D.C., 2011. Defining the public health threat of food fraud. Journal of Food Science 76 (9), R157–R163.

Stader, J., 2015. Saint Lous Survey on Seafood Mislabeling Bonafid. http://www.bonafidcatch.com/assets/bonafid_study2015.pdf.

Stiles, M.L., Kagan, A., Lahr, H.J., Pullekines, E., Walsh, A., 2013. Seafood Sticker Shock: Why You May Be Paying Too Much for Your Fish. From: http://oceana.org/sites/default/files/reports/Oceana_Price_Report.pdf.

Stoeckle, K., Strauss, L., 2008. Students Use DNA Barcodes to Unmask "mislabeled" Fish at Grocery Stores, Restaurants.

Strickland, J., Boyd, T., April 28, 2015. Bait and switch: metro restaurants mislabeling fish. Atlanta Journal Constitution. http://www.ajc.com/news/entertainment/dining/bait-and-switch-metro-restaurants-mislabeling-fish/nk45Q/.

Teletchea, F., Laudet, V., Hänni, C., 2006. Phylogeny of the Gadidae (sensu Stetovidov, 1948) based on their morphology and two mitochondrial genes. Molecular Phylogenetics and Evolution 38, 189–199.

Tennyson, J.M., Winters, K.S., Powell, K., October 6–7, 1997. A Fish by Any Other Name: A Report on Species Substitution. Paper Presented at the 22nd Annual Meeting of Seafood Science the Technology Society of the Americas, Biloxi, Mississippi.

U.S. Food and Drug Administration, 2014b. Fish: What Pregnant Women and Parents Should Know. From: http://www.fda.gov/food/foodborneillnesscontaminants/metals/ucm393070.htm.

U.S. Food and Drug Administration (FDA), 2011. Fish and Fishery Products Hazards and Controls Guidance, fourth ed. Gainesville, FL. http://www.fda.gov/downloads/Food/GuidanceRegulation/UCM252383.pdf.

U.S. Food and Drug Administration (FDA), 2014a. FDA DNA Testing at Wholesale Level to Evaluate Proper Labeling of Seafood Species. http://www.fda.gov/Food/GuidanceRegulation/GuidanceDocumentsRegulatoryInformation/Seafood/ucm419982.htm.

Usydus, Z., Szlinder-Richert, J., Adamczyk, M., Szatkowska, U., 2011. Marine and farmed fish in the Polish market: comparison of the nutritional value. Food Chemistry 126 (1), 78–84.

Vasquez, M., August 23, 2009. Snapper on your plate may be an imposter. The Miami Herald.

Warner, K., Timme, W., Lowell, B., Stiles, M., 2012a. Persistent Seafood Fraud Found in South Florida. From:http://oceana.org/en/news-media/publications/reports/persistent-seafood-fraud-found-in-south-florida.

Warner, K., Timme, W., Lowell, B., Hirshfield, M., 2012b. Widespread Seafood Fraud Found in L.A. From: http://oceana.org/en/news-media/publications/reports/widespread-seafood-fraud-found-in-los-angeles.

Warner, K., Timme, W., Lowell, B., 2012c. Widespread Seafood Fraud Found in New York City. From:http://oceana.org/en/news-media/publications/reports/widespread-seafood-fraud-found-in-new-york-city.

Warner, K., Timme, W., Lowell, B., Hirschfield, M., 2013. Oceana Study Reveals Seafood Fraud Nationwide: Oceana.

Warner, K., Golden, R., Lowell, B., Disla, C., Savitz, J., Hirshfield, M., 2014. Shrimp: oceana reveals misrepresentation of America's favorite seafood. Oceana.

Warner, K., Lowell, B., Disla, C., Ortenzi, K., Savitz, J., Hirshfield, M., 2015. Oceana reveals mislabeling of iconic chesapeake blue crab. In: Oceana.

Warner, K., 2011. Seafood Fraud Found in Boston-Area Supermarkets. From:http://oceana.org/en/news-media/publications/reports/seafood-fraud-found-in-boston-area-supermarkets.

Weaver, K.L., Ivester, P., Chilton, J.A., Wilson, M.D., Pandey, P., Chilton, F.L., 2008. The content of favorable and unfavorable polyunsaturated fatty acids found in commonly eaten fish. Journal of the American Dietetic Association 108, 1178–1185.

Wolf, I., June 11, 2009. Bice Busted: Eateries Caught in Fish Fraud, Chicago Sun-Times. Scripps Howard News Service. http://www.nova.edu/ocean/ghri/forms/news-chicago-sun-times-bice-busted-eateries-caught-in-fish-fraud.pdf.

Wong, E.H.-K., Hanner, R.H., 2008. DNA barcoding detects market substitution in North American seafood. Food Research International 41 (8), 828–837. http://dx.doi.org/10.1016/j.foodres.2008.07.005.

Chapter 2

Seafood Traceability and Consumer Choice

Eric Enno Tamm[1], Laurenne Schiller[2], Robert H. Hanner[3]

[1]This Fish/EcoTrust Canada, Vancouver, BC, Canada; [2]Vancouver Aquarium, Vancouver, BC, Canada; [3]University of Guelph, Guelph, ON, Canada

Traceability has largely been considered a technical requirement for business to meet government regulations regarding food safety, food recalls, and country-of-origin labeling. Consumers have traditionally placed confidence in government regulations and the reputation of consumer brands to ensure products are safe, good quality, responsibly sourced, and properly labeled. However, the last 20 years have seen significant changes in consumer attitudes and trust toward corporate brands and government. DNA-based analysis has played an important role in exposing seafood fraud and mislabeling in the market place and therefore has helped to raise consumer and government awareness about the need for enhanced traceability (Mariani et al., 2014). More recently, there have been technological changes, especially the rise of social and mobile technologies, that are empowering consumers to take a more active role in value chains.

The confluence of these social and technological trends has given rise to new electronic traceability systems that provide consumers with detailed product information on key data elements and critical tracking events that have typically been accessible to only government or business. These new electronic traceability systems are providing consumers with more refined data on their seafood, potentially including where, when, and how it was caught. This new technology enables consumers to retrieve traceability data on a product through an alphanumeric code, bar code, or quick-response (QR) code that can be scanned or traced on a smartphone, tablet, or computer. In this way, the technology is similar to software systems used by international couriers such as FedEx, DHL, or UPS to enable customers to track packages in real time.

This chapter provides a brief history on the evolution of traceability from its inception in the 1960s to today, breaking its development into three distinct periods. The chapter then delves into the most recent period looking at the social, economic, and technological factors that have enabled the rise of consumer-facing traceability in seafood products and surveys the current consumer-facing

Seafood Authenticity and Traceability. http://dx.doi.org/10.1016/B978-0-12-801592-6.00002-4

27

traceability service providers. The chapter concludes by analyzing the demographics, motivations, and response rates of consumers using ThisFish, a traceability system launched by Ecotrust Canada in 2010. This case study provides preliminary insights into the future of consumer choice and seafood traceability.

EVOLUTION OF TRACEABILITY

Modern-food traceability has undergone three major transformations since its inception in the early 1960s. It started as collaboration with Pillsbury, NASA, and the US Army Laboratories to provide safe food for space expeditions (Sperber and Stier, 2009). This early food safety research eventually led to Pillsbury developing a training program for the US Food and Drug Administration titled "Food Safety Through the Hazard Analysis and Critical Control Point System," or HACCP as it is widely known today. This first generation of food traceability was focused on food safety, involved one-up-one-down traceability and was paper based. Although HACCP was eventually imposed through government regulation, companies also saw self-interest in HACCP to mitigate the risk to their trademarks from lapses in food safety that would harm their customers and trigger expensive food recalls, lawsuits, and lost consumer trust. In the seafood industry, HACCP became especially important for live shellfish, such as clams, mussels, and oysters, that are vulnerable to natural and man-made toxins in the environment.

The second generation of seafood traceability was in response to growing consumer distrust in both business and government, and rising environmental awareness. The 1980s saw the start of the first major seafood campaign, which was a boycott of canned tuna from fisheries that harmed dolphins. In 1990, the "dolphin-safe" ecolabel was launched, followed by the first certification, the Marine Stewardship Council (MSC), in 1996 (Jacquet et al., 2009). These market-based certifications were in response to both real and perceived failures on the part of government and business to manage fisheries' sustainably as evidenced by the collapse of Canada's Atlantic cod fishery in 1992.

Besides certifying fisheries as sustainable, MSC included a chain-of-custody certification to trace fish from the certified fishery of origin to the final point of sale. This second generation of traceability was largely driven by market-based certification systems that included chain-of-custody standards. In 2005, the Food and Agriculture Organization published *Guidelines for the Ecolabelling of Fish and Fishery Products from Marine Capture Fisheries* to establish basic requirements for chain of custody and traceability of certified products (Food and Agriculture Organization of the United Nations, 2009). Other certifications followed that were based on the FAO Code of Conduct for Responsible Fisheries (Food and Agriculture Organization of the United Nations, 2011). These certifications claimed that products were from "responsibly managed" fisheries and included Friends of the Sea (2006), the Alaska Responsible Fisheries Management Certification (2011) (The Alaska Responsible Fisheries Management (RFM) Chain of Custody, 2016), and Iceland Responsible Fisheries Certification (2011). New sustainable and responsible certifications were also launched for aquaculture products: the Aquaculture Stewardship Council (2010) and Global Aquaculture Alliance Best Aquaculture Practices standard (2011) (Global Trust Certification, 2016). The International Fishmeal and Fish Oil Organization created a "responsibly sourced" certification in 2013 (IFFO RS CoC Standard, 2016), and in 2014, both Fair Trade USA and TRU-ID Inc. began seafood certification.

These certifications are voluntary, market-based schemes that focus on consumer choice, providing a logo or "trust mark" that is meant to guarantee the authenticity of a social or environmental claim. To ensure that their logos aren't misused, the certifications also include chain-of-custody and traceability requirements. The growth of certifications was not limited to the seafood industry either, but was part of a larger trend which included organic, ethical, socially responsible, and sustainable certifications for forest products, produce, meat, diamonds, coffee, cocoa, textiles, etc.

This second generation of traceability was similar to the first generation of HACCP-focused traceability in that it was still one-up-one-down traceability and was linked to a paper trail of documents in the value chain. However, it differed in that these chain-of-custody certification standards were audited by independent third-party certifying bodies. It should also be noted that during this period a growing number of companies began using information technology in the form of enterprise resource planning (ERP) and accounting software to manage their production, sales, and logistics data. Data began its migration from paper to the digital realm.

The third generation of traceability began in the mid-2000s and was driven by a confluence of three factors: major changes in technology generally and information technology specifically, growing market concerns about illegal fishing and seafood fraud, and changing consumer attitudes and preferences. This period would begin to see the use of electronic cloud-based software that provides full-chain traceability with the ability for consumers to trace the origins of their seafood through websites and smartphone apps.

TECHNOLOGICAL REVOLUTION

Three major changes in technology have facilitated the development of the third generation of seafood traceability. First, the growth of the Internet and specifically cloud-based enterprise software has meant that a vast amount of data are now stored electronically and are accessible in real time via computers and smartphones. Internet usage rates have risen to 78.3 per 100 inhabitants in 2014 in developed countries and to 31.2 in developing countries (International Telecommunications Union, 2015). This cloud-based computing has enabled seafood businesses to share data with one another and with consumers.

Second, there has been an explosion in the use of mobile phones and smartphones, which enable consumers to access an unprecedented level of information about consumer goods in the palm of their hands. Mobile-cellular subscriptions in the world went from 33.9 per 100 inhabitants in 2005 to 95.5 in 2014 (International Telecommunications Union, 2015), and mobile-broadband subscriptions have risen to 83.7 per 100 inhabitants in developed countries in the same period. Mobile technologies have enabled businesses to upload data to cloud-based software from remote locations and have enabled consumers to retrieve data on products at the point of sale.

Third, social media technologies have reached widespread penetration rates with 71% of online adults in the United States using Facebook, 28% using LinkedIn, and 28% using Pinterest (Duggan et al., 2015). Consumers are also actively rating and commenting on products on e-commerce sites such as Amazon, or posting reviews on Facebook, Twitter, Yelp, Urbanspoon, and many other websites. The rise of social media, the Internet and mobile technologies has empowered consumers to actively engage value chains through product reviews; crowdfunding; microlending; direct online sales; and product testing, tracing, and tracking.

Research from the National Restaurant Association (NRA) shows that a majority of Americans have used touch-screen ordering, smartphone apps, and mobile payments as part of their dining experience (National Restaurant Association, 2013). The survey found that about 63% of adults used restaurant-related technology. The numbers are even higher for younger adults: about 70% of consumers aged 18–34 found restaurants on smartphones and half used computers to order food or make reservations. "Following wider societal technology trends, we're seeing that younger consumers are much more likely to interact with restaurants on their smartphones than older adults," said Hudson Riehle, the NRA's senior vice president of Research and Knowledge. "However, there is a substantial number of older consumers who say they would use smartphone apps for certain things, like looking up directions and finding nutrition information."

While technology has enabled greater levels of consumer engagement in the value chain, it has been media investigations and nongovernmental organization (NGO) campaigns on illegal fishing and seafood fraud that have provided the motivation to consumers and policy makers to demand greater levels of traceability.

ILLEGAL FISHING AND SEAFOOD FRAUD CAMPAIGNS

In the past decade, there has been rising awareness about seafood from illegal, unreported, or unregulated (IUU) fisheries entering value chains and fraud as a result of seafood mislabeling. The awareness has largely been driven by investigative reporting and activist NGOs. One of the primary tools used to uncover seafood fraud has been DNA-based analysis of product samples purchased in the marketplace, both in restaurants and retail outlets.

One of the earliest investigations was by the *New York Times* in 2005 in which reporters collected samples of wild salmon from eight stores in New

York City. DNA analysis subsequently found that many of the samples were actually cheaper farmed salmon (Burros, 2005). Other journalistic investigations followed including the Canadian Broadcasting Corporation's Marketplace investigative report undertaken in collaboration with the Biodiversity Institute of Ontario (BIO) in 2010 which found in one of every five fish they purchased in Canadian stores was mislabeled according to DNA testing (CBC Radio Canada, 2010). The following year, in 2011, the *Boston Globe* completed a 5-month investigation in collaboration with BIO whereby reporters collected seafood samples from 134 restaurants, grocery stores, and seafood markets (The Boston Globe, 2016). The results showed that 87 of 183 products, or 48%, were mislabeled. The investigation focused on tuna and snapper, the species most frequently mislabeled. *Consumer Reports* conducted their own survey that same year, by DNA testing in 190 pieces of seafood bought at retail stores and restaurants in New York, New Jersey, and Connecticut (Consumer Reports Magazine, 2011); and from 2010 to 2012, Oceana, a global conservation NGO, conducted a series of surveys collecting more than 1200 seafood samples from 674 retail outlets in 21 US states to determine whether DNA testing (again, in collaboration with BIO and others) revealed mislabeling. Testing found that 33% of the analyzed samples were mislabeled according to US Food and Drug Administration guidelines (Oceana, 2013).

Though helping to raise awareness about seafood fraud, DNA-based analysis of seafood samples has its limitations in terms of authenticating the provenance of a product. Although DNA analysis can authenticate the species and even a genetically distinct population within a species, it cannot authenticate that the seafood came from a legal fishery or that it was harvested using a sustainable method. A number of product attributes cannot be authenticated using DNA-based analysis. Yet the inability to properly document species identity as revealed by DNA testing leads to questions concerning the veracity of other reported attributes.

A study published in *Marine Policy* found that global seafood supply chains have become so opaque and convoluted "that consumers and vendors of fish are generally unaware of the role they play in buying and selling illegally caught products" (Pramod et al., 2014). The researchers studied 180 information sources and conducted 41 interviews that focused on the top 10 seafood-importing countries to the United States. They found that between 20% and 32% ($1.3 to $2.1 billion USD) of wild-caught seafood imported into the United States is estimated to be illegally harvested. Some of the worst countries for illegal seafood are China, where the researchers estimate 45–70% of salmon and 30–45% of pollock is illegally caught, often from Russian fisheries. Illegal and unreported fishing is hidden by its nature as products enter complex and diverse supply chains that may include trans-shipments at sea, landing, and transit between countries for various stages of processing, including the division and combination of lots.

Even if fish is properly labeled and is harvested in a legal fishery, human and labor rights may have been violated in its harvesting, transportation, and production. Investigations by the Associated Press (McDowell et al., 2015) and *Guardian* (Hodal et al., 2014) have uncovered human rights violations and slavery involving migrant workers in fisheries in Thailand and Indonesia. FishWise and the Environmental Justice Foundation (2014) have also recently conducted studies into human trafficking, slavery, and human rights violations in fisheries. The FishWise white paper documents pervasive human trafficking, forced labor, child labor, and egregious health and safety violations in the seafood industry, especially in the developing world. The problem has been exacerbated by corrupt officials, fishing vessels in international waters flying "flags of convenience" to get around stricter regulations of other nations, the globalization of supply chains, IUU fishing, and the increased number and mobility of migrant workers from poor countries. Victims on fishing vessels are typically boys and men between the age of 15 and 50. On shore, victims typically are female fish plant workers.

Improved transparency and traceability in the fishing industry are commonly called for by those concerned with improving seafarer well-being and achieving seafood sustainability. Implementing full traceability at every step in the supply chain to trace receipt, processing, and shipping of seafood is the first step in eliminating IUU fishing, human rights abuses, and seafood fraud. Full supply chain traceability should make it easier to identify and remove seafood associated with these concerns from supply chains and, ideally, prevent it from entering the supply chain initially (FishWise, 2013).

These high-profile investigations have led to rising awareness among consumers, industry personnel and policy-makers about the need for traceability in seafood supply chains. In 2010, the European Union introduced Catch Certificates (Regulation No. 1005/2008) to trace imported seafood back to fishing vessels. The certificate requires information about the product's catch vessel, transport vessel, scientific name, and FAO catch area, among other data. In 2014, the European Union also introduced new seafood labeling laws (European Union regulation No. 1379/2013) requiring businesses to provide details on catch methods and harvesting areas to consumers. The US Congress also introduced the Safety and Fraud Enforcement for Seafood Act in 2013 (Library of Congress, 2013) and in 2014 a Presidential Task Force on IUU Fishing and Seafood Fraud recommended the creation of a risk-based traceability program to track seafood from harvest to entry into US commerce, although it curiously failed to include a role for DNA-based verification testing (Office of the Federal Register, 2014).

Media investigations and NGO campaigns targeting seafood have been effective at setting the agenda and raising awareness. However, these seafood campaigns also occurred during a period of heightened concern about meat mislabeling in the United Kingdom after the "horsegate" scandal, tainted produce scandals in the United States and the global outbreak of disease among farm animals including avian flu, swine flu, and mad cow disease. All these issues have helped to change perceptions and raise concern among consumers about the provenance of food, making it a mainstream issue among diverse consumer groups today.

CHANGING CONSUMER PREFERENCES

Outside the food industry, a number of general market trends are increasing the importance of traceability to consumer choice. First, the Internet revolution has empowered consumers to actively search for product information and to even create content through product reviews. Hence the Internet is now shaping nearly every aspect of a consumer's purchasing decision (Business Development Bank of Canada (BDC), 2013). Nine of 10 consumers, according to the survey of 1023 Canadian consumers, claim to use their smartphone for preshopping activities, and more than two in five consumers use their smartphone to find promotional offers, one-third check product reviews, and a similar proportion verify product availability in local stores. Consumers are embracing new behaviors largely empowered by the revolution in information technology.

Second, aging populations in the West are heightening concerns about food safety, health, and nutrition. As populations grow grayer, they also tend to grow more conscientious about what they eat. Older consumers are more likely to look for products and services to help them maintain and improve their health. Studies have shown that health concerns are the greatest driver of increased seafood consumption, especially among older adults (Perishables Group, 2011). Seafood has a reputation for being a low-fat protein source with the added benefits of omega-3 fatty acids and other nutrients. Recent food mislabeling and tainted produce scandals have heightened consumer concern about food safety.

Third, consumers are increasingly using their wallets to influence companies to adopt more responsible practices. In fact, close to 6 in 10 Canadians consider themselves ethical consumers. People want products that reflect their own values. These values could include environmental protection, animal welfare, child-free labor, fair trade, and buying local. For example, two-thirds of Canadians claim to have made an effort to buy local or Canadian-made products, while nearly three-quarters of consumers say they will pay more for locally produced food, in part because most believe it is fresher and tastier. Similarly, most Canadians say they would pay more for a restaurant meal if all the ingredients were produced locally.

A market trend report by Innova Market Insights identified traceability as one of the top 10 trends in the food industry for 2014, suggesting businesses are looking for solutions to industry challenges and changing consumer preferences. According to the Innova survey, global product launch activity featuring the word "origin" for claims purposes increased by 45% for the first half of 2013 compared with the second half of 2012. Innova also found that about two-thirds of US survey respondents preferred to buy food if they know the origins and 29% find local claims to be very or extremely important when making a purchasing decision (Food Business News, 2014). Origin claims are valuable beyond Western consumers or "locavores." An online survey of Chinese con- sumers in Beijing and Shanghai, who purchase seafood at least once a month to eat at home, found product preference changed with origin claims. Almost three-quarters (73%) of survey respondents indicated that Japan has the best reputation for wild salmon compared with 39% for the United States and 27% for Russia (Chinese Seafood Consumers, 2012).

SEAFOOD ECOLABELS IN NORTH AMERICA AND THE CONSERVATION ALLIANCE FOR SEAFOOD SOLUTIONS

The late 1990s and early 2000s saw an increase in consumer awareness initia- tives in sustainable seafood. In 1996, the largest global ecocertification body, the MSC, was formed in London, United Kingdom, as a partnership between the World Wide Fund for Nature and Unilever. This organization contracts indepen- dent audits for fisheries seeking ecocertification and certifies fisheries that meet its specific environmental and management standards. All fisheries that meet the MSC's standards are marked with the MSC ecolabel. A similar organization for aquaculture, the Aquaculture Stewardship Council (ASC) was established in 2010. Although not directly affiliated with the MSC, the ASC conducts similar audits based on the impacts of both land-based and ocean-based aquaculture practices around the world, and certified operations bear the ASC ecolabel.

In North America, many aquariums took the lead in incorporating private, consumer-facing seafood recommendation programs into their outreach and education strategies. Leaders include the Shedd Aquarium (Right Bite pro- gram), New England Aquarium, and Monterey Bay Aquarium (Seafood Watch program) in the United States and the Vancouver Aquarium (Ocean Wise pro- gram) in Canada. Unlike the MSC, these organizations do not audit or certify fisheries, but rather work largely with retailers and restaurants to encourage seafood retailers, restaurants, and consumers to source and eat products deemed "sustainable." For the most part, ecological sustainability is the key focus of their recommendations and a specific set of publicly available Capture Fish- ery and Aquaculture Criteria published by Monterey Bay Aquarium's Seafood Watch is used to conduct science-based assessments of various seafood opera- tions. Products scoring highly are those that have the fewest impacts on the fish stock and surrounding ecosystem.

In Canada, the two main seafood recommendation organizations are the Vancouver Aquarium's Ocean Wise program, and SeaChoice (a collaboration of the David Suzuki Foundation, Canada Parks and Wilderness Society, Living Oceans Society, and the Ecology Action Centre that was established in 2006). Both programs have consumer-facing ecologos; however, SeaChoice is focused on large retail chains, while Ocean Wise works mostly with the foodservice industry and regional retailers. Ocean Wise began in 2005 with 16 restaurant partners and was inspired by Chef Robert Clark's passion for ocean health and choosing seafood responsibly. It has grown into a national sustainable seafood program engaging chefs, seafood suppliers/producers, chain and independent restaurants, and retailers.

In addition to consumer-facing programs, a variety of other North American NGOs working throughout the supply chain on traceability and sustainability issues have also been established in the last decade. To tackle problems in a more coordinated, holistic manner, the Conservation Alliance for Seafood Solutions was subsequently established in 2008. At present, 16 different US- and Canada-based NGOs form the basis of the Conservation Alliance network, collaborating to help seafood industry business partners (i.e., fisheries, suppliers, retailers, restaurants, etc.) meet their commitments to sourcing and selling sustainable seafood. Together, they have established a Common Vision which identifies six critical areas, where seafood companies can demonstrate environmental leadership and take action to ensure a sustainable seafood supply. These include:

- Make a Commitment
- Collect Data
- Buy Environmentally Responsible Seafood
- Be Transparent
- Educate
- Support Reform

More information on the Conservation Alliance for Sustainable Seafood and the specific eNGO's that participate can be found at www.solutionsforseafood. org. Consolidated support for change is an important step toward achieving a more ethical and sustainable seafood supply, but process verification is essential to traceability (Olsen and Borit, 2013) and to ensure such programs achieve their intended results.

TRACEABILITY AND CONSUMER CHOICE

Governments, businesses, and consumers have identified traceability as a valuable product attribute. Its value proposition can vary depending on the fishery, country of origin, product type, and complexity of the supply chain. Traceability supports a variety of important initiatives concerning food safety, mislabeling and fraud, IUU fishing, human rights, fishery improvement projects, marketing and promotion, and meeting sustainability commitments (Boyle, 2012). Many of these uses focus on building greater consumer trust in product claims. The potential consumer benefits of traceability include:

- *Reducing Mislabeling and Seafood Fraud.* Traceability creates trusted product information so that customers can be guaranteed that they are getting what they paid for. It can reduce the amount of product from IUU fishing in the marketplace and make it more difficult for businesses to mislabel products once they are in the supply chain.

- *Improved Safety and Quality.* Traceability can improve accountability so that fish harvesters, processors, and distributors handle product more responsibly, since it can be traced back to them. This accountability and transparency should improve quality over time.
- *Improved Grading.* Since fish harvesters and seafood suppliers likely want to trace their best quality products, there is an incentive to improve product grading. It will also be more difficult for fish harvesters and suppliers to cheat in grading, since wrongly graded product can be traced back to them.
- *Better Market Differentiation.* In commodity industries, businesses struggle to differentiate their products beyond price. Traceability allows businesses to tell more detailed stories about their products and to validate their claims. This could prove especially beneficial for businesses wanting to market products as being from a particular country or region or harvested with a unique or sustainable method.

The past decade has seen the development of a number of electronic traceability systems that include features to engage the public to trace their products. The online engagement typically consists of mobile apps, websites, mobile sites, scannable QR codes, or traceable alphanumeric codes. There are over a dozen nonprofits and companies that offer consumer-facing traceability services, typically in the form of software as a service (SaaS), while several companies have also developed their own proprietary systems. Some examples are covered in Table 2.1.

Although the opportunities for consumers to engage in food traceability have increased since the mid-2000s, there is currently little data or research concerning which consumers are most likely to engage in tracing their products, what motivates them to trace, and what are their response rates when offered the opportunity to trace their products.

In September 2007, YottaMark conducted a national survey of 2700 US households on food traceability (Grant, 2007). It is the parent company of HarvestMark, software that allows consumers to trace the origins of their fresh produce. The survey respondents were mostly mothers who did the shopping for their families. Overall, the survey found that consumers trust retailers, but their confidence in the safety of the food they purchase had been seriously shaken by tainted produce scandals. The survey found that 90% of consumers trusted large retailers and 82% trusted specialty stores to sell safe produce. However, fewer than 30% gave their unqualified trust; most shoppers were only "somewhat" confident. The survey also found that confidence with half the respondents was declining. Fully 83% of consumers said they would use a website at home to trace produce at the unit level, at least occasionally. Fewer than 10% said they would not use an online traceability tool. About 87% said they would use an in-store kiosk, at least on some items. Some 28% said they would use it on all produce.

Asked about the information they wanted on their food, 74% said they "definitely" wanted information on food safety, recalls and alerts. Harvest

TABLE 2.1 Companies and NGOs Offering Consumer-Facing Traceability Services

Company	Product Name	Website	Description
Traceability Service Providers			
BackTracker	FishTale	backtrackerinc.com and fishtale.co.nz	Consumers access product information through webpages and scannable QR codes. Companies can customize the information seen by consumers.
Computer Associates Inc.	Seasoft ERP and SeaTrace app	www.caisoft.com	Consumers use a browser-based lot tracking and traceability smartphone app that provides detailed information about the seafood they are purchasing. The SeaTrace app utilizes QR codes on product labels.
GS1 Germany	fTrace	www.ftrace.com	fTrace is a website, mobile site, and mobile app that allows consumers to trace a code to retrieve data on where the product comes from, when and how it was produced, and info about quality and recipes. Uses both scannable QR and traceable alphanumeric codes.
Gulf Wild	TransparenSea	www.gulfwild.com	Consumers use a 9- or 10-digit unique code on gill tags that can be traced online to retrieve product details. Uses both scannable QR and traceable alphanumeric codes.
ScoringAg Inc.	Traceback	traceback.com	Consumers use a traceable alphanumeric code on product to trace product details online.
ShellCatch	ShellCatch	www.shellcatch.com	Information on traceability and environmental practices is delivered to consumers through scannable QR and traceable alphanumeric codes.
Fish Trax	Fish Trax Marketplace	marketplace.fishtrax.org	Consumers can use a scannable QR or traceable alphanumeric code to retrieve data on the harvesting and production of their seafood. Uses a website for tracing.
Ecotrust Canada	ThisFish	thisfish.info	Consumers use a website and mobile site to trace the origins of their seafood and connect to the harvester who caught it. Uses scannable QR and traceable alphanumeric codes.

Continued

TABLE 2.1 Companies and NGOs Offering Consumer-Facing Traceability Services—cont'd

Company	Product Name	Website	Description
Trace and Trust	Trace and Trust	traceandtrust.com	Consumers us a website and mobile app to see details of seafood deliveries to restaurant and retailer location. The system uses hyperlinks to share data on each delivery from fish harvesters and fish farms to retail outlets and restaurants.
Trace Register	Trace Register	www.traceregister.com	A customizable marketing module allows restaurants and retailers to share traceability data with their customers via a mobile app. Uses both scannable QR codes and traceable alphanumeric codes.
Traceall Global	Traceall Global	www.traceallglobal.com	The system creates customized reports and brand-specific web portals to share traceability information with consumers.
TraceTracker	Food Explorer	tracetracker.com	Consumers can search for the ingredients, product history, nutritional information, or environmental records of individual products or product lines through websites, mobile devices, and kiosk.
Proprietary Traceability Systems			
Red's Best	Red's Best	www.redsbest.com	Traceability software tracks product from boat to plate through the use of scannable QR codes connected to data on website and mobile site.
John West	John West	www.john-west.co.uk	Company-specific software that allows consumers to use numeric barcodes and can codes to trace the origins of their seafood on canned product through website.
Bumble Bee	Bumble Bee	www.bumblebee.com	Announcement of consumer-focused traceability on cans to be implemented in the summer of 2015.

ERP, Enterprise resource planning; *QR,* quick response.

date (57%), where the product was grown (46%), nutritional information (43%), ability to rate quality (30%), links to organic certification/history (26%), and specific harvest data such as farmer's name (23%) were also ranked as important attributes. The survey found that 85% of consumers responded that they would probably or definitely choose a traceable product over one that was not traceable, all other things being equal. It also found that about 55% were willing to pay between 1% and 5% premium on the retail price. These findings are notable, however, YottaMark's research described consumer intent, which can vary significantly when compared with actual shopper behavior.

In 2014, ThisFish, a consumer-facing traceability system and initiative of Vancouver-based nonprofit Ecotrust Canada, conducted an online survey of consumers who traced seafood on its system and also sent a comment to their fish harvester. The online survey was sent to 1118 consumers via email with 302 responding, a response rate of 27%. About 93% of the responses came from Canada, 5% from the United States, and 2% from Europe. The survey sample represented the most engaged consumers, since they both traced their seafood and also sent a message to their fish harvester via ThisFish.

To benchmark the consumers who traced their fish, ThisFish compared its survey results against a survey of 3004 Americans conducted in 2012 by National Public Radio (NPR) on sustainable seafood (Table 2.2). Even accounting for the difference in nationality and survey timing, the comparison showed

TABLE 2.2 Comparison of Consumer Responses to Sustainable Seafood Surveys Conducted by ThisFish and NPR

	ThisFish Survey (%)	NPR Survey (%)
Frequency of purchasing fish (>4 times per month)	68	30
Importance of sustainably caught fish (very important)	73	34
Importance of high quality (very important)	91	61
Likelihood of buying sustainably labeled seafood (very likely)	63	24
Confidence in sustainability label (very confident)	26	14
Willingness to pay premium for sustainable seafood (>10%)	58	25

TABLE 2.3 Consumer Responses to Questions Concerning Consumer Motivation to Use ThisFish to Trace Their Seafood

What Motivated You to Trace Your Seafood Using ThisFish? I Was Interested in... (%)	
Where my seafood was harvested	85
Who harvested my seafood	77
When my seafood was harvested	66
How my seafood was harvested	59
The novelty of tracing my seafood	56
Food safety and health	28
Ecorating and certification	23
More info on fish species	19
Nutritional info	8

that those consumers who traced their fish using ThisFish were more than twice as likely to eat fish more than four times a month and were twice as likely to consider sustainability very important. They also exhibited twice the likelihood to pay a premium for sustainable food.

The survey responses received by ThisFish also asked respondents about their motivation (Table 2.3) to use a code to trace their seafood. About 85% of respondents wanted to know where their seafood came from and 77% wanted to know the fish harvester who caught it. In written comments to this question, many respondents cited concerns about contaminants, pollution, and processing in foreign countries such as China. Many expressed a desire to eat local.

Consumers also had very high expectations regarding the characteristics of traceable seafood (Table 2.4). They expected traceable seafood to be more likely properly labeled, fresher, more sustainable, higher quality, local, independently harvested, and more expensive.

Although the consumer survey by ThisFish is only one case study, the results suggest that consumers who trace their seafood value quality and sustainability, and demonstrate a willingness to pay premium prices. They are primarily motivated by the opportunity to receive information on the who, what, when, where, and how their seafood was harvested. They also have high expectations that traceable seafood is better quality, more sustainable and more expensive. The data seems to suggest that many tech-savvy consumers seek a sense of connection to their food supply and because younger consumers are tech savvy and represent the fastest growing segment of the market, this trend is likely to increase.

TABLE 2.4 Consumer Expectations Concerning Traceable Seafood Tagged via ThisFish

In Your Opinion, If a Seafood Product is Traceable, How Likely Is It to Be...	Likely (%)	Very Likely (%)
Properly labeled	45	48
Fresher	43	44
More sustainable	49	40
Higher quality	50	41
From an independent fish harvester	45	40
Local	40	32
More expensive	66	13

CONSUMER TRACEABILITY: RESPONSE RATES

Given the opportunity to trace a seafood product, what percentage of the population would actually scan a QR code or trace a code on a mobile device or computer? How do you benchmark the "tracing rate" to determine if the consumer response was high or low? Currently, there is a dearth of data to answer these questions.

The traceability system supported by ThisFish collects data on consumer tracing through web analytics. Each time a code is traced by a consumer, the system collects web-analytics data on the tracing activity, including the IP address, time, date, and location of IP address. These data are then displayed to fish harvesters and seafood businesses on private, password-protected dashboards via their user accounts on the system. The dashboard feature provides market feedback to the seafood industry.

ThisFish conducted two preliminary case studies to benchmark consumer tracing rates. The first case study involved traceable lobsters from Nova Scotia. Lobstermen tagged 275,698 lobsters with a uniquely coded plastic tag branded with the ThisFish logo on one side and the phrase "Discover the story of your seafood. Trace this code... www.thisfish.info" on the other side. The lobsters were sold to the live market in North America, Europe, and Asia. The lobster sample was weighted for 20% loss due to tag breakage, mortality, and some lobsters being sent to processing plants because of poor quality. In these cases, the traceable tag would never make it into the hands of consumers, chefs, or retailers. The tracing activity showed that about 66% of traces came from Canada, 14% from the United States, 2% from the Netherlands, and 18% from the rest of world. The sample was also weighted for the fact that 87% of the North American population has access to the Internet. This case study focused on a global supply chain with a live product.

The second case study involved a small integrated seafood business in the Netherlands that consisted of both a fishing boat and retail shop. The fish harvester tagged some 6829 mullet and European sea bass with uniquely coded tags. The fish was sold fresh to local markets in the Netherlands within several days of harvesting. The sample was weighted for only 1% loss due to breakage and against 93% of the Dutch population having access to the Internet. It must also be mentioned that neither case study involved point-of-sale marketing or marketing campaigns explaining to consumers what traceability is, or that they could trace their seafood. The marketing was passive whereby consumer would discover the traceable attribute of the product only by reading the ThisFish tag affixed to the product.

ThisFish calculated the tracing rate as a percentage of traces relative to the number of unique codes on lobsters and individual fish. These tracing rates were then compared with other response rates for physical and digital marketing tactics. For example, online advertising uses "click through rates" which are the proportion of clicks relative to views or impressions of ads, opened emails, or sponsored content. For direct mail, response rates are calculated based on the number of letters, flyers, catalogs, etc., that are distributed relative to the number of consumers responding to this printed material. In 2012, Nellymoser, a digital marketing company, also analyzed response rates for QR codes and other interactive apps printed in magazine ads. Comparative response rates (Table 2.5) for online marketing, direct marketing, QR-coded ads, and ThisFish traceability suggest that consumers exhibit a relatively strong interest in food attributes.

Overall, the tracing rates for ThisFish's cases studies were relatively high, ranging from 10.4% to 12.9%, compared with other marketing tools. The next highest response rate was QR-coded magazine adds at 6.4%. The case studies provide some cautious optimism regarding consumer engagement in traceability. However, tracing rates may vary depending on point-of-sale marketing, the type of product and market demographics. Tracing rates may also be high due to the novelty of the ThisFish tag, which would diminish over time.

CONCLUSION

Seafood traceability has undergone three fundamental changes since its modern inception in the late 1960s. These changes have primarily been driven by the globalization of supply chains, rising consumer concern about social and environmental issues, and the revolution in biotechnology and information technology. Seafood traceability was initially driven by food safety concerns in the 1970s. In the 1990s, new ecolabels imposed chain-of-custody and traceability requirements on the seafood industry as a means to ensure that their trademarks were not being misused in the marketplace. Although not infallible, chain-of-custody certification provided greater assurance for consumers that a product was from

TABLE 2.5 Performance Rates for Online Marketing, Direct Marketing, QR-Coded Ads and ThisFish Traceability (Salesforce, 2013; Matus and Carver, 2012)

Marketing Response Rates[a,b]		
Marketing Tool	Metric	Performance (%)
Online ads	Click through rate	0.10
Facebook ads	Click through rate	0.12
Email	Click through rate	0.12
Paid Search[c]	Click through rate	0.22
Promoted Tweets	Click through rate	2.00
Direct mail (letter)	Response rate	3.40
Direct mail (catalog)	Response rate	4.26
Direct mail (overall)	Response rate	4.40
Magazine QR/app	Response rate	6.40
ThisFish Nova Scotia Lobster	Tracing rate	10.41%
ThisFish Netherlands Fish	Tracing rate	12.90%

QR, Quick response.
[a]Salesforce, 2013. The Facebook Ads Benchmark Report. www.social.com.
[b]Matus, R., Carver, A., 2012. Scan Response Rates in National Magazines based on Nellymoser Companion Apps. Nellymoser.
[c]Matus, R., Carver, A., 2012. Scan Response Rates in National Magazines based on Nellymoser Companion Apps. Nellymoser.

a certified fishery. Technology has helped to drive the most recent generation of consumer-focused traceability. First, DNA-based analysis empowered news media and environmental NGOs to expose widespread mislabeling and seafood fraud in the market place. Investigative reporting and NGO activism demonstrated the need for greater levels of authenticity and traceability—especially electronic traceability—in supply chains. At the same time, changes in information technology—including the rise of the Internet, cloud-based computing and smartphones—enabled businesses to collect, store, and share traceability information directly with consumers. The evolution of seafood traceability has been toward increasing and empowering consumer choice, but it may ultimately harness the potential for consumers to access DNA-based authentication themselves directly through their smartphones (e.g., LifeScanner DNA identification kit: http://www.scientificamerican.com/citizen-science/lifescanner/).

Current research and data on consumer engagement with traceability systems is still in its early stages. There is very little published data on consumer choice and traceability or peer-reviewed scholarship. As a case study, ThisFish has illuminated the consumer motivations for tracing seafood products and provided a few data points for benchmarking tracing rates against other marketing response rates. Further research needs to be conducted to validate these preliminary findings.

Authenticity and traceability are crucial for market-driven forces to impact meaningful change to the negative aspects of the global seafood industry. Seafood fraud dilutes the impact of consumer choice to support ecolabels, traceability tools, and the underlying concepts (e.g., sustainability) they promote. Ultimately without some form of verification testing to underpin ecolabeling and traceability schemas, they may not achieve their aims and could eventually be perceived by consumers as ineffectual, or worse. However, we see cause for optimism given increasing consumer awareness of the economic, environmental, health-related, and social issues at play. This awareness, combined with the purchasing power of the new generation of tech-savvy consumers, suggests that change is not only possible, but also likely, particularly because of the decreasing cost and increasing accessibility of this technology for consumers.

REFERENCES

Boyle, M., August 2012. Without a trace: an updated summary of traceability efforts in the seafood industry. FishWise.

Business Development Bank of Canada (BDC), 2013. Mapping Your Future Growth: Five Game-Changing Consumer Trends.

Burros, M., April 10, 2005. Stores Say Wild Salmon, but Test Say Farm Bred. New York Times. http://www.nytimes.com/2005/04/10/dining/10salmon.html?_r=0.

CBC Radio Canada, 2010. Something's Fishy & Busted: Easyhome. http://www.cbc.ca/marketplace/episodes/2010-episodes/somethings-fishy.

Chinese Seafood Consumers, 2012. A Survey of Retail-Purchasing Behaviors. Seafood Source. Portlant, ME. pp. 16 and 18.

Consumer Reports Magazine, 2011. Mystery Fish: The Label Said Red Snapper, the Lab Said Baloney. http://www.consumerreports.org/cro/magazine-archive/2011/december/food/fake-fish/overview/index.htm.

Duggan, M., et al., 2015. Social Media Update 2014. Pew Research Center. http://www.pewinternet.org/2015/01/09/social-media-update-2014/.

Environmental Justice Foundation, March 4, 2014. Slavery at Sea: The Continued Plight of Trafficked Migrants in Thailand's Fishing Industry. http://www.ejfoundation.org/node/1062.

FishWise, November 2013. Trafficked: Human Rights Abuses in the Seafood Industry, p. 14.

Food and Agriculture Organization of the United Nations, 2009. Guidelines for the Ecolabelling of Fish and Fishery Products From Marine Capture Fisheries. Rome http://www.fao.org/docrep/012/i1119t/i1119t00.htm.

Food and Agriculture Organization of the United Nations, 2011. Code of Conduct for Responsible Fisheries. Rome http://www.fao.org/docrep/013/i1900e/i1900e00.htm.

Food Business News, 2014. Ten Food Trends Unveiled at I.F.T. 2014. http://www.foodbusinessnews.net/articles/news_home/Consumer_Trends/2014/06/Ten_food_trends_unveiled_at_IF.aspx?ID=%7BC8BEEAF6-9CD1-401B-9A19-972D66B243C4%7D&cck=1.

Friend of the Sea, 2006. http://www.friendofthesea.org/about-us.asp?ID=2.

Global Trust Certificatin, 2016. http://www.gtcert.com/layout/content_details.cfm/ck/31/ctk/35/k/118.

Grant, E., October 1, 2007. "Consumers Show Clear Preference for Traceable Foods. Research Pater #3/07. YottaMark.

Hodal, K., Kelly, C., Lawrence, F., June 10, 2014. Asian Slave Labour Producing Proawns for Suplermarkets in US, UK. The Guardian. http://www.theguardian.com/global-development/2014/jun/10/supermarket-prawns-thailand-produced-slave-labour.

I.F.F.O.R.S. CoC Standard, 2016. http://www.iffo.net/iffo-rs-coc-standard.

International Telecommunications Union. Statistics. http://www.itu.int/en/ITU-D/Statistics/Pages/stat/default.aspx.

Jacquet, J., et al., 2009. Conserving wild fish in a sea of market-based efforts. Oryx, The International Journal of Conservation 1–12.

Library of Congress, 2013. S:520 – 113th Congress (2013–2014). https://www.congress.gov/bill/113th-congress/senate-bill/520.

Matus, R., Carver, A., 2012. Scan Response Rates in National Magazines Based on Nellymoser Companion Apps. Nellymoser.

McDowell, R., Mason, M., Mendoza, M., March 25, 2015. Hundreds Forced to Work as Slaves to Catch Seafood for Global Supply. Associated Press. The Globe and Mail. http://www.theglobeandmail.com/news/world/men-forced-to-work-as-slaves-to-catch-seafood-for-global-supply/article23609283/.

Mariani, S., Ellis, J., O'Reilly, A., Bréchon, A.L., Sacchi, C., Miller, D.D., September/October 2014. Mass media influence and the regulation of illegal practices in the seafood market. Conservation Letters. 7 (5), 478–483. http://onlinelibrary.wiley.com/doi/10.1111/conl.12085/full.

National Restaurant Association, 2013. New Research Shows Consumers Want a Side of Technology With Their Meals. http://www.restaurant.org/News-Research/News/New-research-shows-consumers-want-a-side-of-techno.

Oceana, 2013. Oceana Study Reveals Seafood Fraud Nationwide". http://oceana.org/reports/oceana-study-reveals-seafood-fraud-nationwide.

Office of the Federal Register, 2014. Traceability: Create a Risk-Based Traceability Program to Track Seafood From Harvest to Entry into US Commerce. https://www.federalregister.gov/articles/2014/12/18/2014-29628/recommendations-of-the-presidential-task-force-on-combating-illegal-unreported-and-unregulated#h-19.

Olsen, P., Borit, M., 2013. How to define traceability. Trends in Food Science & Technology. 29, 1–9. http://doi.org/10.1016/j.tifs.2012.10.003.

Perishables Group, 2011 Consumer Trends, www.perishablesgroup.com.

Pramod, G., Nakamura, K., Pitcher, T., Delagran, L., 2014. Estimates of illegal and unreported fish in seafood imports to the USA. Marine Policy 48, 102–113.

Salesforce, 2013. The Facebook Ads Benchmark Report. www.social.com.

Sperber, W.H., Stier, R.F., December 2009. Happy 50th Birthday to HACCP: Retrospective and Prospective. Food Safety Magazine 42–46.

The Alaska Responsible Fisheries Management (RFM) Chain of Custody, 2016. http://certification.alaskaseafood.org/chain-of-custody.

The Boston Globe, 2016. Globe Investigation Finds Widespread Seafood Mislabeling. http://www.boston.com/business/specials/fish_testing/.

Chapter 3

Regulatory Frameworks for Seafood Authenticity and Traceability

Johann Hofherr[1], Jann Martinsohn[1], Donna Cawthorn[2], Barbara Rasco[3], Amanda M. Naaum[4]

[1]European Commission, Joint Research Centre (JRC); [2]University of Stellenbosch, Stellenbosch, South Africa; [3]Washington State University of Idaho, Pullman, WA, United States; [4]University of Guelph, Guelph, ON, Canada

The level of development and enforcement of traceability systems and regulations on food adulteration and misbranding are critical in addressing seafood mislabeling and also affect the integration of DNA-based testing as a tool to manage the associated health, economic, and environmental concerns. The regulatory framework varies between regulatory and governmental bodies in different countries or unions. For example, the 21 nations of the Organization for Economic Cooperation and Development (OECD) widely differ in terms of their mandatory traceability regulations at the national level, regulations and required documentation for imported foods, and sophistication and accessibility of documentation and labeling at the consumer level. In a work by Charlesbois et al. (2014), a ranking of countries was conducted based upon an assessment of traceability programs (Table 3.1).

European Union (EU) regulations are arguably the most comprehensive and progressive in addressing concerns of seafood fraud and fisheries management. Developing countries, on the other hand, may have more fragmented regulations, or lack the necessary resources to enforce those that may be in place. In this chapter, we review the existing regulations and challenges in effective government monitoring of the seafood supply chain using examples from these two jurisdictions. We focus on the background, existing regulations and the role of voluntary certification systems and also discuss the application of DNA-based technology and the importance of better standardization and support for seafood authenticity and traceability globally. The differences between these established and emerging traceability systems and regulations for seafood provides

Seafood Authenticity and Traceability. http://dx.doi.org/10.1016/B978-0-12-801592-6.00003-6
47

TABLE 3.1 Rating of Sophistication of Food Traceability System

Progressive		Moderate	Regressive	No Data
Austria	Germany	Australia	China	Russian Federation
Belgium	Ireland	Brazil		
Denmark	Italy	Canada		
Finland	The Netherlands	Japan		
France	Sweden	New Zealand		
Norway	United Kingdom	United States		
Switzerland				

Progressive, Country has specific traceability regulations for all commodities;
Moderate, Country with some but less broad or stringent regulations;
Regressive, Countries in the developmental stage with traceability regulations, whether government or industry driven.

a snapshot of two different extremes in global management of this industry. We conclude the chapter with an overview of new regulations in the United States, the second largest importer of seafood in the world, and how these will affect the global trade of seafood.

TRACEABILITY DEFINED

Traceability is the ability to access all properties of a food product and its ingredients, including the ability to track credence attributes, such as country of origin, source, and production methods. Ultimate traceability systems would be able to provide information regarding these properties of a food product or ingredient in all of its forms, at all points across the value chain. Traceability is not only required at the product batch level, but also at the facility level, and at least one step forward and one step backward in the supply chain. These systems provide the basis for recall programs, which are efficient if it is possible to link traceability programs across a supply chain. Also in the context of recalls, traceability provides the basis for mitigating damages from an intentional contamination incident. The capability to facilitate product tracking both backward and forward across supply chains is critical and for complex supply chains, such as those associated with seafood (Fig. 3.1), they can be difficult to both establish and maintain. In seafood trade, traceability plays a major role in tracking products in international markets, in terms of providing a way to track harvest area, identity of fish harvested, and handling and storage conditions. Improving

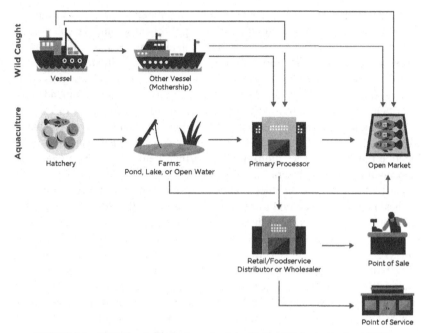

FIGURE 3.1 Example of a seafood supply chain (National Fisheries Institute, 2015).

inventory control is possibly the greatest advantage of enhanced traceability from a business perspective as it provides better monitoring of perishable goods and their transit to markets, reduces instances of counterfeiting, and helps control shrinkage and other forms of product loss.

An ideal traceability system in the seafood supply chain would provide the ability to track the fish from harvest across the whole value chain, as well as ensure that the products flowing through the chain are safe, legally caught, and honestly labeled. This requires tracking from vessels/farms, through slaughter, portioning, and processing and through the full range of distribution channels to the consumer. Analytical testing at various stages is critical to ensure that authentic, safe products are being transported through the supply chain and to limit opportunities for undetected fraudulent substitutions or additions.

Definitions of traceability vary, but an effort to provide a harmonized definition (Olsen and Borit, 2013) suggests that the traceability system must:

- provide access to the information on all properties of seafood product;
- provide access to the information on ingredients of seafood product in all its forms;
- facilitate traceability both backward and forward;
- be based on systematic recordings of these properties.

Further important attributes would include: reliability, a means of seamless data sharing among multiple sources in the supply chain, common technological architecture to facilitate data sharing, low-tech alternatives, ease of use, low or no associated use cost, and support for multicultural and multilanguage settings. In order to implement such a system the Institute of Food Technologists (IFT) seafood industry advisory panel (2014 meeting) suggested a recommended list of critical tracking events (CTE), including: transformation input, transformation output, transportation, and depletion, including consumption and disposal. Key data elements to support the CTE events are: event owner, date/time, event location (GLN, global location), trading partner, item ID (GTIN, global trade item number), lot/batch/serial number, quantity, unit of measure, activity ID, and activity type. Though these recommendations present useful guidelines for best practices in seafood traceability, the reality is that few systems, either mandatory or voluntary, exist that track this level of information. Also missing is the integrated use of analytical testing necessary to support traceability systems. Verification testing is a critical, but sometimes overlooked, component of traceability. Although testing facilities and tools exist to analyze many types of potential food adulteration, access to these facilities may be limited, funding to support testing may be insufficient, or regulations on testing may be unclear or nonexistent. The ability to provide reliable testing varies significantly between different jurisdictions, as the remainder of this chapter illustrates.

FAO CODE FOR RESPONSIBLE FISHERIES

Traceability in the food supply chain is increasingly becoming a requirement in major fish-importing countries and for many internationally operating retailers. It can safeguard public health and demonstrate that fish has been caught legally from a sustainably managed fishery or produced in an approved aquaculture facility. To provide guidance, the FAO describes with its Code of Conduct for Responsible Fisheries (FAO, 1995) best practices for certification of products and processes and for ensuring that labels on fish products are accurate and verifiable. After almost two decades since its adoption, the Code of Conduct for Responsible Fisheries (the Code) remains key to achieving sustainable fisheries and aquaculture. It provides a framework, and its implementation is steered by 4 international plans of action (IPOAs), 2 strategies, and 28 technical guidelines, which have evolved to embrace the ecosystem approach. Most countries, including the EU, have fisheries policy and legislation that are consistent with the Code, while other countries have plans to align them. Globally, the priority for implementation is the establishment of responsible fisheries with due consideration of relevant biological, technical, economic, social, environmental, and commercial aspects. Members have reported progress on various aspects of the Code including establishment of systems to control fisheries operations, developing food safety and quality assurance systems, establishment of mitigation measures for postharvest losses, and development and implementation of national plans to combat illegal, unreported, and unregulated (IUU) fishing and curtail fishing capacity. Several regional fishery bodies (RFBs) have implemented management measures to ensure sustainable fisheries and protect endangered species. The 2012 independent evaluation of FAO's support for the implementation of the Code was positive but called for more strategic and prioritized development and support, improved outreach, closer articulation between normative and operational work, and more attention to the human dimensions. For developing nations, such as South Africa, the FAO Code serves as a model for generating national legislation for seafood traceability, management, and regulation.

THE REGULATORY FRAMEWORK FOR SEAFOOD AUTHENTICITY AND TRACEABILITY IN THE EUROPEAN UNION

Within the EU, the Common Fisheries Policy (CFP) (European Parliament and Council, 2013), sets the regulatory framework for the exploitation of living marine resources. The CFP aims to ensure that fishing and aquaculture are environmentally, economically, and socially sustainable and that they provide a source of healthy food for EU citizens. Its goal is to foster a dynamic and profitable fisheries and aquaculture industry and to ensure good socioeconomic conditions in coastal communities. In pursuing the previously mentioned aims, a number of rules and laws constitute a legislative framework, implemented under the remit of the CFP, which also address the fight against illegal fishing, food safety and hygiene, consumer protection, level playing field for producers, and environmental concerns. In this context, traceability and authentication are indispensable pillars of control and enforcement, with monitoring and verification schemes helping to enhance compliance with existing rules.

The following section provides an overview of the history of EU food and feed authentication and traceability, discusses existing EU legislation underpinning seafood authenticity and traceability, and reflects to what extent DNA technology is considered in and endorsed by existing EU legislation.

Food Traceability in the European Union

Traceability of agricultural products, particularly of livestock, goes back a long way in human history (Blancou, 2001). With an early focus on claiming ownership and origin, the scope of application grew to include trade monitoring, animal disease containment, and consumer protection. In Europe, the Common Agricultural Policy (CAP) was created in 1962 to ensure food security, affordable prices for consumers, and acceptable living standards for farmers (European Commission and Directorate-General for Agriculture and Rural Development, 2012). This encompassing policy framework together with the establishment of a common internal market, enabling the free circulation of livestock, food and feed products between EU member states, and a constant rise in volume and intricacy of international trade, led to the need to design and implement increasingly sophisticated traceability and authentication schemes to accommodate the challenges emerging from enhanced production and liberated trade.

Some of the first community-wide rules for the internal market concerning the traceability of livestock go back to 1964 (Council Directive 64/432/EEC of June 26, 1964). The target at that time was primarily to control animal health risks, which could impact the intra-community trade in cattle and pigs. In the early 2000s, in the wake of the crisis caused by bovine spongiform encephalopathy (BSE), which was followed by a major foot-and-mouth disease (FMD) outbreak and after some health risks linked to the consumption of meat, milk, and eggs that originated from feed contaminations with dioxins, traceability schemes for livestock were further enhanced. Stricter and more encompassing traceability rules were not only aimed at enhancing consumer protection against health risks and fraud, but also to ensure a high level of animal health in the EU and to minimize the risk of trade impediments caused by epidemics and health concerns.

To enhance consumer protection against health risks and incorrectly labeled food and to rebuild consumer confidence lost during the disease outbreaks mentioned earlier, in 2002 the "Farm to Fork" (European Commission, 2004a), or "Stable to Table" (Seimenis and Economides, 2002), traceability principle covering the entire production chain was implemented in the EU feed and food sector. This principle is anchored in Regulation 178/2002 (European Parliament and the Council, 2002), the general food law of the EU, which renders traceability compulsory for all food and feed businesses. Food and feed operators are required to implement a comprehensive traceability system that enables the identification of origin and destination of their products and moreover allows the rapid provision of this information to the competent management authorities. Within the EU food safety policy, traceability serves as a risk management tool helping food business operators and regulators to identify, and, if necessary, withdraw or recall products, which constitute a public health risk. The requirement of traceability also applies to fishery products, from the net to the plate, although some differences to other food commodities exist in this area.

In the general food law, traceability is defined as "the ability to trace and follow a food, feed, food-producing animal or substance intended to be or expected to be incorporated into a food or feed, through all stages of production, processing and distribution." Traceability should enable producers as well as authorities to keep the production and supply chain under control by establishing a connection between all steps along the chain. This should also enable the monitoring of ingredient(s) used and incorporated along the chain and the verification that information given for a product is correct.

This implies the need for unambiguous identification of livestock and food products and the tracking of their transit through the supply chain from origin to retail. Moreover tracking must be possible in both directions, downstream and upstream the supply chain. Downstream (forward) tracking is mainly a tool for planning and safe sourcing of the primary materials needed for processing purposes, while upstream (backward) tracking is applied to find the source of a product in question. To achieve a reliable identification of products in a traceability scheme, according to EU law, business operators are required to document for each consignment the name and address of suppliers and customers, the nature of the product and date of delivery. In addition, operators are encouraged to keep information on volume or quantity of the product, the batch number (if applicable), and a more detailed description of the product, for example, whether it is raw or processed.

The European Commission has published guidelines addressed at all stakeholders in the food chain to support the comprehension, correct interpretation, and application of requirements emerging from the general food law (http:// ec.europa.eu/food/food/foodlaw/guidance/docs/guidance_rev_8_en.pdf).

In addition to the general traceability requirements, sector-specific legislation applies to some categories of food products, such as vegetables, olive oil, fruit, beef, honey, and fish, in order to provide consumers with validated information on origin and authenticity. Meanwhile, specific information requirements are extended to a range of other food products under the regulation regarding food information for consumers (European Parliament and the Council, 2011).

While traditionally, traceability schemes are sustained by marks/tags and written documents and certificates, advanced technologies are increasingly used to enhance such schemes, the most prominent being radio-frequency identification (RFID). RFID is an automatic identification method, enabling electronic retrieval and tracking of the carrier of identification code of RFID tags or transponders incorporated into a tag (Domdouzis et al., 2007) and entered progressively into the voluntary and compulsory identification of animals; in the EU, RFID has become the official means of identification in small ruminants (Council, 2003) and equidae (European Commission, 2008). Moreover, in some EU member states, it is also used voluntarily for identification of cattle. For animals, RFID is based on the international standards ISO 11784 and ISO 11785. Interestingly, an RFID-enabled traceability system has also been proposed for the supply chain of live fish (Hsu et al., 2008). While a major advantage of RFID is its robustness against tampering, it is costly as compared to traditional approaches, and information output depends on the availability of an electronic reader.

Besides the identification of a product, documentation; certification and the proper transfer of information are essential parts in chain traceability as, for example, established under the EU veterinary and public health rules. Preserving the integrity of traceability schemes is a complex and challenging endeavor, and ideally dedicated bodies or institutions are established to monitor and control such schemes. Within the EU a trans-European network, TRACES (TRAde Control and Expert System), for veterinary health, has been established, which monitors, notifies, and certifies imports, exports, and trade of animals and animal products. Industry and competent authorities all over the world can use this web-based network free of charge to trace back and forth animal and animal products. TRACES allows users to produce and exchange certificates, related to animals, animal products, by-products, semen, and embryos, in 22 official EU languages, based upon the latest consolidated model. While not currently feasible to be directly applied to seafood, this approach illustrates one concept of a traceability standard for animal products.

The traceability requirements for genetically modified organisms (GMOs) in the EU represent an example of how DNA testing can be integrated into a traceability scheme. The attitude toward GMOs differs substantially between Europe and the United States, rendering it a highly interesting socioeconomic and policy-related subject. In 2001 Lynch and Vogel analyzed the divergent approaches of the EU and the United States toward the introduction and marketing of genetically modified (GM) foods and seeds, leading to considerable challenges to trans-Atlantic and international trade (Lynch and Vogel, 2001). The prevalent cautious and reserved attitude toward GMOs in the EU resulted in a focus on traceability schemes suitable for GMOs and GM products intended for release into the environment or for entering into feed and food. The highly elaborate monitoring and traceability scheme for GMOs in the EU is reviewed briefly here, as it might serve as a paradigm for traceability of food commodities in general, including fish and fish products.

For GMOs the EU follows the precautionary approach and puts emphasis on consumer information, by implementing stringent approval, labeling, and traceability standards on any food produced from or derived from GM ingredients. This implies that new technologies and strategies for detecting and tracing the presence and amount of GMOs in agricultural products had to be established. Independent control methods, supporting documentation and certification, capable of identification, and, ideally, quantification, had to be found for GMOs to ensure that the content of such organism in a product could be traced and that the accuracy of labels could be tested. It is obvious that the very nature of GMOs and GM products suggest applying DNA-based testing approaches to support traceability of GMOs and GMs (Miraglia et al., 2004). EU legislation for GMOs regulates issues concerning environmental aspects, food and feed safety, procedures for commercialization, and labeling provisions (European Parliament and the Council, 1997; European Parliament and Council, 2001, 2003a).

Traceability and labeling requirements are laid down in Regulation (EC) 1830/2003 (European Parliament and Council, 2003b) while methods of sampling and controls of non-authorized GM materials in feed are set out in Regulation (EU) 619/2011 (European Commission, 2011a).

If companies apply in the EU for the environmental or market release of a GMO after control and approval of the European Food Safety Authority (EFSA), subsequent control and monitoring falls under the remit of the European Network of GMO Laboratories, which was inaugurated in 2002 (ENGL—http://engl.jrc.ec.europa.eu). It constitutes EU experts working on the development, harmonization, and standardization of methods for detection, identification, verification, and quantification of GMOs or derived products, such as seeds, grains, food, feed, and environmental samples.

Detection and quantification of GMOs in food and feed based on validated protocols are indispensable, since, according to EU legislation, a content of more than 0.9% GMO in any authorized product (technical zero level=0.1% threshold if non-authorized) must be indicated on labels (European Parliament and Council, 2003a). An additional challenge arises since each genetic modification requires a specific detection protocol ("event-specific detection") To ensure that monitoring and control of each new GMO is possible, ENGL since 2004 provides assistance to the Community Reference Laboratory (CRL— http://gmo-crl.jrc.ec.europa.eu/) for GM Food and Feed, particularly with respect to the validation of analytical methods for the event-specific quantification of GMOs that are under marketing approval. Details of these tasks are anchored in EU legislation on GM Food and Feed (European Parliament and Council, 2003a). The identification of new GM products during the application process and later surveillance is simplified by the compulsory insertion of a nucleotide-sequence (undisclosed information) in the DNA by the producer and the communication of this information to the CRL (European Commission, 2004b, 2013). Due to the conditions and requirements imposed by EU legislation regarding DNA analysis, and its supporting infrastructure, GMO monitoring and traceability is clearly advanced as illustrated by the established group of certified laboratories ENGL. This might to some extent help to design DNA-based traceability schemes for fisheries and aquaculture products. In addition, with recent approval of a GMO salmon by the FDA in the United States (http://www.fda.gov/AnimalVeterinary/DevelopmentApprovalProcess/GeneticEngineering/GeneticallyEngineeredAnimals/ucm280853.htm), continued DNA testing will be increasingly relevant to traceability and identification of GMO in aquaculture products.

Current European Union Regulations for Seafood Authenticity and Traceability

As previously mentioned, the CFP aims to ensure that EU fishing and aquaculture are environmentally, economically, and socially sustainable and that they provide a source of healthy food for its citizens. To ensure compliance with rules set under the CFP, a community control system was established:

1. the Control Regulation 1224/2009 (Council, 2009) ensuring compliance with the rules of the Common Fisheries Policy;
2. the IUU Regulation (EC) No. 1005/2008 (European Council, 2008a) concerning illegal, unreported, and unregulated fishing;
3. the Fishing Authorisation Regulation (EC) No. 1006/2008 (European Council, 2008b) for fishing activities of Community fishing vessels outside Community waters and the access of third country vessels to Community waters.

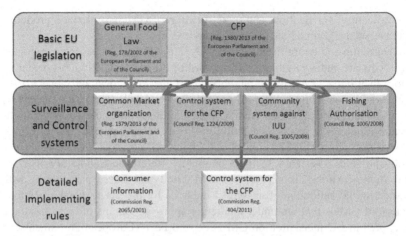

FIGURE 3.2   Schematic overview of EU legislation with relevance for authenticity and traceability in fisheries and aquaculture products.

This full system entered into force on January 1, 2010. The three pillars are complementary and will ensure that there is no discrimination between the EC and third country fisheries (Fig. 3.2).

The CFP Control Regulation specifies control requirements, based on not only established but also new technologies, such as vessel detection systems and automatic identification systems. The legislation opens up for further electronic monitoring devices and interestingly also emphasizes in Article 13 traceability tools, such as genetic techniques. Some of the most important elements of the control regimes the member states have to implement are:

- Log book for vessels of 10-m length or more, submit landing and transshipment declarations (electronically for vessels of 12 m or more and prenotification of landings).
- Transshipments only in designated ports and monitoring transshipments by competent authorities.
- The whole chain of production and marketing should be covered by a coherent traceability system, complementary to the basic rules of the food law (Regulation 178/2002—see previous section). In the interest of consumers, fishery and aquaculture products that are marketed within the EU, irrespective of their origin, shall contain information concerning the commercial designation, production method, and the catch area at each stage of the marketing as stipulated in Regulation 1379/2013 (Regulation (EU) No 1379/2013 on the Common Organisation, 2013) and 2065/2001 (European Commission, 2001).

Regulation 1224/2009 also reiterates the general food law requirements for fisheries and aquaculture products, stating that all lots have to be traceable at all stages of production, processing, and distribution, from catching or harvesting to the retail stage (Art. 58):

- Fisheries and aquaculture products placed on the market or likely to be placed on the market in the Community have to be adequately labeled to ensure the traceability of each lot.
- Lots of fisheries and aquaculture products may only be merged or split after first sale if it is possible to trace them back to catching or harvesting stage.
- The Member States have to ensure that the operators have systems and procedures in place to identify any operator from whom they have been supplied with lots of fisheries and aquaculture products and to whom these products have been supplied. This information has to be made available to the competent authorities on demand.
- The minimum labeling and information requirements for all lots of fisheries sand aquaculture products have to contain:
  - Identification number of each lot
  - External identification number and name of the fishing vessel or the name of the aquaculture production unit
  - FAO alpha-3 code of each species
  - Date of catches or date of production
  - Quantities of each species in kilograms expressed in net weight, or when appropriate, the number of individuals
  - Information to the consumer as referred to below (Art. 35 of Reg. 1379/2013).
- Some of the information on the lots does not apply to fisheries and aquaculture products imported into the EU with catch certificates submitted in accordance with the IUU Regulation (European Council, 2008a). Member States may also exempt small quantities of products sold directly from fishing vessels to consumers (max value 50€ per day) from the labeling requirements.
- Further rules apply for first sale (first marketed or registered at auction centers or at registered buyers or producer organizations—Art. 59), weighing of fisheries products (Art. 60–61), completion and submission of sales notes (Art. 62–65), take-over declarations (Art. 66–67), transport documents (Art. 68), monitoring and surveillance (Art. 69–81).

With Regulation 404/2011 (European Commission, 2011b) the European Commission laid down detailed rules on the implementation of the control systems pursuant to the CFP Control Regulation 1224/2009, providing rules for the vessel monitoring system, satellite-tracking devices, frequency of data transmission, fishing logbooks, transshipment declarations, landing declaration in paper or electronic formats, and traceability. While, as mentioned earlier, DNA analysis is mentioned in Article 13 of the CFP Control Regulation, its application in the context of traceability, control, and enforcement is currently suggested to be tested in pilot studies carried out by member states or the European Commission and therefore is not strictly required. However, EU projects, such as LabelFish (labelfish.eu), are beginning a framework for incorporating DNA into seafood traceability. This project, currently focused on the Atlantic Area, is focused on genetic control of seafood labeling and traceability, and could potentially see expansion to other parts of the EU.

EU legislation relevant for seafood traceability also includes Regulation 1379/2013 on the common organization of the markets in fishery and aquaculture products. It stipulates, in detail, the labeling requirements for fishery and aquaculture products for sale to the final consumer/caterer:

Mandatory labeling requirements (Art. 35):

- Commercial designation of the species and its scientific name in accordance with FishBase (http://www.fishbase.org/) or the ASFIS database from FAO (http://www.fao.org/fishery/collection/asfis/en);
- Production specification: "caught" or "caught in freshwater" or "farmed";
- Origin specification through indication of the area where the product was caught (for marine fisheries: sub-area or division listed fin FAO fishing areas, for Inland fisheries: name of water body), or in the case of aquaculture products where it was farmed (Member State or third country in which product reached more than half of its final weight or stayed for more than half of the rearing period, or in case of shellfish underwent a final rearing or cultivation stage of at least six months)
- Category of fishing gear used for capture (Annex III to the regulation);
- Whether the product has been defrosted and the date of minimum durability, where appropriate;
- Competent authorities responsible for monitoring and enforcing rules, making full use of available technology, including DNA-testing in order to deter operators from falsely labeling catches.

Additional voluntary information if verifiable (Art. 39) can be added, such as date of catch or harvest, date of landing, information on the port at which the product was landed, more detailed information on the fishing gear, flag state of the vessel, information on environmental, ethical, or social nature, information on production techniques and practices, nutritional content of the product, and QR code on mandatory information. Member states are obliged to undertake checks to ensure compliance with the previously mentioned labeling rules. These checks may take place at all marketing stages and during transport (Art. 45).

More specific traceability rules for fisheries and aquaculture are laid down in the Commission Regulation 404/2011 (European Commission, 2011b), which emphasizes that reliable authenticity and traceability are key components of the legislative framework and the rules established under the CFP.

In fisheries and aquaculture, authenticity and authentication includes the ability to identify the species and to discriminate and trace the origin of fish (i.e., from which population or stock they are derived). While genetic species identification, even on highly processed products, is meanwhile fairly established (albeit a coherent EU-wide approach and infrastructure is still lacking), genetic origin assignment depends on the establishment of a robust baseline, that is, the identification of (genetic) populations[1] of a species across an area of interest. This depends on a considerable investment in a fundamental research approach as was done for EU waters in the FP7 project FishPopTrace (https://fishpoptrace.jrc.ec.europa.eu/) (Nielsen et al., 2012) and for aquaculture, relating to genetic traceability of wild and farmed marine fish, in AquaTrace (https://aquatrace.eu/). The intention of these research projects is to establish genetic markers as discriminating parameters between populations or wild and farmed fish as reference points in a traceability system.

Voluntary Certification Systems and Ecolabeling in the European Union

In the light of the need to move toward a more sustainable and responsible exploitation of marine natural resources, the interest in the certification of wild capture fisheries and aquaculture production systems, practices, processes, and products is increasing. Certification of fisheries and aquaculture products being caught or produced in a sustainable way is also endorsed for fishery and aquaculture products by the EU. As stated in the legislation on a Common Market Organization in fisheries and aquaculture products, the Commission will submit, after consulting member states and stakeholders, a feasibility report on options for an ecolabel scheme for fishery and aquaculture products, in particular on establishing such a scheme on a Union-wide basis and on setting minimum requirements for the use by member states of a Union ecolabel (Art. 13 of Reg. 1379/2013 (Regulation (EU) No 1379/2013 on the Common Organisation, 2013)). Based on a feasibility study (Sengstschmid et al., 2011) and the opinion of the EU Ecolabeling Board[2] in 2011, the Commission is at the time of writing not intending to develop Ecolabel criteria for food and feed products. Though not required, voluntary certification of seafood products is widely available in the EU. This may include ecolabeling like that available from MSC. It may also include certification of DNA authenticity, like that available from TRU-ID (www.tru-id.ca).

DNA Technology and Application in the Context of Legislative Frameworks in the European Union

As illustrated earlier, authenticity and traceability of food commodities, including seafood, are strongly endorsed by the current existing policy and legislative

---

1. The terms "stock" and populations have been discussed extensively in Waples, R.S., Gaggiotti, O.E., 2006. What is a population? An empirical evaluation of some genetic methods for identifying the number of gene pools and their degree of connectivity. Molecular Ecology 15, 1419.
2. http://ec.europa.eu/environment/ecolabel/documents/EUEB_position_on_food_final.pdf.

framework in the EU. For seafood, clearly DNA analysis can greatly contribute to establishing independent control means supporting the authentication of information required under the remit of traceability schemes. One example of this is the Norwegian minke whale register, which allows the genetic identification of each individual whale from legal harvest (Glover et al., 2012). Although this level of identification is currently difficult to scale across the whole seafood industry, this shows the potential power of DNA-based traceability. As delineated in other chapters of this book, for species identification of fishes, DNA analysis is well established, but origin assignment will require further fundamental research, though this area is strongly supported by the developments in genetics and genomics (e.g., in geographical provenance, population assignment, or verification of food origin (Ogden and Linacre, 2015; New Analytical Approaches, 2013)). For both, moving toward a coherent and standardized approach across the EU, venues to create a supportive and cost-efficient infrastructure, well imbedded in a legislative framework, should be explored. Facilities exist across the EU with the capacity to carry out DNA testing, so access is not a critical limiting factor. As discussed earlier, the agricultural sector and also GMOs can serve at least partly as a paradigm for developing a more standardized approach, and large-scale research projects are being undertaken within the EU to continue to improve the capacity for both species and population identification of seafood. In general, it is important for the management of the CFP[1] to be guided by principles of good governance. Those principles include decision making based on best available scientific advice, broad stakeholder involvement, and a long-term perspective.

THE REGULATORY FRAMEWORK FOR SEAFOOD AUTHENTICITY AND TRACEABILITY IN DEVELOPING COUNTRIES: A SOUTH AFRICAN PERSPECTIVE

The global fisheries sector has changed dramatically over the past three to four decades from an industry that was long dominated by the developed world. During this time, developing countries have substantially expanded their share in the total demand, production, and trade of fishery products, largely as a result of their increased participation in the global economy (OECD, 2010). In spite of growing concerns on the state of many marine resources, the worldwide consumption of seafood has doubled since the 1970s. The developing world has been responsible for about 90% of this growth, mainly fueled by escalating human populations, rising incomes, and urbanization (Delgado et al., 2003). To meet the burgeoning demand for fish, developing countries have drastically increased their overall production (Fig. 3.3), outpacing that of developed countries since the mid-1980s and today supplying over 80% of the world's seafood (FAO, 2014a). An important catalyst for the shift in capture fisheries production from developed to developing nations was the establishment of 200-mile Exclusive Economic Zones (EEZs) in 1982 (UNCLOS, 1982), within which coastal

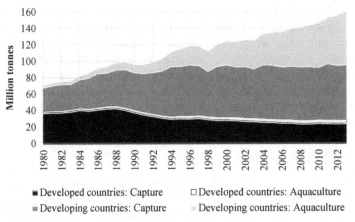

FIGURE 3.3 Increasing participation of developing countries in global seafood production (capture and aquaculture). *Source: FAO, Fishery Statistics.*

nations were able to claim exclusive fishing rights and exclude some developed-country fleets (Pauly, 2009). At the same time, developing countries were expanding their own fishing fleets, while developed countries were contracting theirs (Delgado et al., 2003). The rapid development of aquaculture has further played an integral role in bolstering fish production from developing countries (Fig. 3.1), with the latter now contributing a staggering 94% of the world's farmed seafood products (FAO, 2014a). The value of seafood production in the developing world, however, goes far beyond its direct impact as food. The net exports of fishery products from developing countries reached US $35.3 billion in 2012 and surpassed the monetary revenue accrued from many other agricultural exports, including that of coffee, tea, and cocoa combined (FAO, 2014b). Aside from supplying a crucial source of export earnings, the fisheries sector is also a significant provider of employment in developing regions.

Notwithstanding their substantial contributions to the world's seafood supply, developing countries continue to face immense challenges both at sea and in terms of trade. In particular, weak governance and the limited capacity to effectively "police" their vast ocean expanses put developing countries at considerable risk for illegal, unreported, and unregulated (IUU) fishing by both their own vessels and those from distant water fishing nations (FAO, 2014b; Agnew et al., 2009). In West Africa alone, IUU fishing accounts for up to 40% of the total fish caught (Agnew et al., 2009), which consequently jeopardizes marine ecosystems, socioeconomic development, and food security. Moreover, the deliberate mislabeling of seafood provides an avenue for unscrupulous operators to launder such illegal products into legitimate markets (Pramod et al., 2014). In recognition of the aforementioned problems, more stringent import controls, traceability standards, and seafood labeling requirements have been implemented in parts of the developed world (e.g., in the EU) to prevent illegal and mislabeled products from entering the market. The following uses South Africa as a case study to illustrate the challenges and opportunities encountered by developing countries in setting similar standards and ensuring the legality, authenticity, and traceability of domestically consumed and export-destined seafood products, specifically in the face of fragmented regulatory frameworks, financial limitations, and human-resource constraints. Although a South African perspective is provided, many key issues detailed in the text extend to other developing countries.

The South African Fisheries Sector

South Africa is a nation largely defined by its productive oceans and rich diversity of marine resources, which in turn support a range of industrial, small-scale and subsistence fishermen. The EEZ surrounding continental South Africa has an area of over 1 million km^2 and the coastline extends some 3650 km from Namibia in the west, fed by the cold Benguela current, to Mozambique in the east, fed by the warm Agulhas current (Griffiths et al., 2010). Intense upwelling of nutrient-rich waters along the west coast contributes to high biological productivity, forming the basis for valuable industrial fisheries. Small-scale, subsistence, and recreational fishing is more concentrated on the east coast, which has greater species diversity but less biomass. The country has a small, underdeveloped aquaculture industry, with the few existing aquaculture operations mainly rearing mussels, oysters, prawns, and abalone (Griffiths et al., 2010).

South Africa's domestic marine harvest has averaged over 680,000 tonnes (live weight) per annum over the last decade (FAO, 2014a). Local seafood consumption amounts to 313,000 tonnes per annum (about 6.25 kg/capita/annum), of which about half is imported (Burgener et al., 2014). On the other hand, about 178,000 tonnes of seafood was exported from South Africa in 2010 (FAO, 2014a). While clearly an important player on the African continent in terms of both fisheries production and trade, South Africa has neither escaped the wrath of overexploitation, nor has it evaded the burden of corruption and illegal seafood trade (Hauck and Kroese, 2006). Similar to the global trend of overexploited marine resources, many of the region's wild fish stocks are considered depleted or heavily depleted, predominantly within the inshore zone (DAFF, 2014).

Exposure of Seafood Market Fraud in South Africa

Seafood fraud is globally pervasive and undermines endeavors to achieve sustainable fisheries and transparent seafood trade. Although deceptive marketing of seafood products has been suspected in South Africa for several decades (Smith and Smith, 1966), little was done prior to 2009 to elucidate its true prevalence. A market-based study conducted in 2009/2010 first revealed the failure of many packaged fish products in South African retail outlets to comply with applicable labeling regulations, as well as the vulnerability of consumers to be misled due to the lack of information provided by seafood vendors on the identity, origin, and sustainability of the products on sale (Cawthorn et al., 2011a). Over the past five years, three independent DNA-based studies have been conducted to evaluate the extent of fish mislabeling on the South African market (Cawthorn et al., 2012a, 2015; Von Der Heyden et al., 2010). The earliest two of these studies documented disturbing mislabeling rates of 21–50% (Cawthorn et al., 2012a; Von Der Heyden et al., 2010), both of which cited confused seafood naming practices as a potential source of certain mislabeling incidents and emphasized the inadequacy of existing labeling regulations, traceability systems, and governmental monitoring in addressing these issues. Very few changes transpired, however, until these research findings were televised on a local current affairs journalism program, sparking outrage among consumers and raising industry awareness on the prevalence and risks associated with seafood fraud (Barendse and Francis, 2015). Perhaps partially as a result of this raised awareness, the mislabeling study by Cawthorn et al. (2015) showed slight improvements in the transparency of local seafood marketing (18% mislabeling), but still pointed to a clear problem requiring resolution.

Current Regulatory Framework for the Control of Seafood Trade in South Africa

In South Africa, three national governmental departments (some with subdirectorates or branches) are directly involved in the regulation and control of seafood products, namely the Department of Agriculture, Forestry, and Fisheries (DAFF), the Department of Health (DoH), and the Department of Trade and Industry (the dti) (Table 3.2). The DAFF has the overall responsibility of ensuring the long-term sustainability and orderly access to exploitation of living marine resources, although its jurisdiction also extends to the issuing of permits for seafood product importation. The activities of the DoH are devoted to controlling the safety aspects (not the quality) of locally produced and imported foodstuffs that are intended for the domestic market, including processed food products but excluding raw agricultural commodities (Chanda et al., 2010). The dti regulates the production and trade of all products on the domestic market, under which three agencies are housed that have relevance to seafood trade: the South African Bureau of Standards (SABS), the National Regulator for Compulsory Specifications (NRCS), and the National Consumer Commission (NCC). The SABS is in charge of developing South African National Standards (SANS), the NRCS is tasked with upholding the safety and quality of certain domestically produced, imported, and exported seafood products by ensuring compliance with a set of Compulsory Specifications that are based on SANS, and the NCC is responsible for safeguarding consumer rights. Two further regulatory entities that indirectly influence seafood trade through the setting and levying of import/export duties or rates according to "tariff headings" applied

TABLE 3.2 Summary of the Key Regulatory Bodies, Responsibilities, Regulations, and Policies That Presently Apply to the Trade, Labeling, and Traceability of Seafood Products in South Africa

Department/ Agency	Directorate/ Branch	Responsibilities	Associated Regulations and Specific Provisions Relating to Seafood Authenticity or Traceability
Department of Agriculture, Forestry and Fisheries (DAFF) www.daff.gov.za	**Fisheries management**	• To ensure sustainable management of marine fisheries and orderly access to exploitation (including commercial fishing rights allocations, setting allowable catches/effort, monitoring, control, and surveillance) in accordance with the: **Marine Living Resources Act (MLRA) (Act No. 18 of 1998)**[a] • Issuing permits for operation of fish processing establishments.Issuing permits for fish product importation, in accordance with import control permits required by ITAC.	
Department of Health (DoH) www.health.gov.za	**Food control**	To ensure the safety and compositional standards of processed, domestically produced foodstuffs (including seafood) and to regulate the labeling of processed, pre-packaged foodstuffs in accordance with the: **Foodstuffs, Cosmetics and Disinfectants Act (Act No. 54 of 1972)**[a]	**Regulations relating to the Labelling and Advertising of Foodstuffs (R. 146/2010)**[a] Mandatory information (among other) required on labels of pre-packaged foodstuffs: • name/true description of product that does not mislead/confuse consumer; • name and address of manufacturer, importer or seller; • ingredient list in defined format; • country of origin/processing; • batch number that is identifiable and traceable.
Department of Trade and Industry (the dti) www.dti.gov.za	**South African Bureau of Standards (SABS)** www.sabs.co.za [Standards-setting arm] ⬌	To develop, promote, and maintain South African National Standards (SANS), to promote the quality of goods and services, and to render conformity assessments, in accordance with the: **Standards Act, 2008 (Act No. 8 of 2008)**[a]	**Compulsory Specifications that apply to seafood include:** • canned fish, marine molluscs, and crustaceans (VC 8014:2004)[a] • frozen fish, marine molluscs, and derived products (VC 8017:2003)[a] • frozen rock lobsters and derived products (VC 8020:2003)[a] • frozen shrimps (prawns), langoustines, and crabs (VC 8031:1987)[a] • smoked snoek (VC 8021: 1974)[a] • live aquacultured abalone (VC 9001: 2012)[a]

Organization	Purpose / Act	Details
National Regulator for Compulsory Specifications (NRCS) www.nrcs.org.za [Regulatory arm]	To ensure the safety and quality of specific domestically produced, imported, and exported seafood products by administering and maintaining **Compulsory Specifications**[b] and by implementing regulatory and compliance systems for compulsory specifications, in accordance with the: **National Regulator for Compulsory Specifications Act (Act No. 5 of 2008)**[a]	These compulsory specifications generally require the following information (among other) to appear on relevant product labels: • name[c]/true description of contents that does not mislead/confuse consumer; • name and address of manufacturer, producer, proprietor, or controlling company; • ingredient list in defined format; • country of origin; • date of manufacture; • identification of factory where packed
National Consumer Commission (NCC) www.thencc.gov.za	To implement the **Consumer Protection Act (Act No. 68 of 2008)**[a] which aims to safeguard consumer rights and promote a "fair, accessible, and sustainable marketplace for consumer products and services"	**Consumer Protection Act (Act No. 68 of 2008), part E:** "A producer, importer, distributor, retailer, or service provider must not market any goods… 1. in a manner that is reasonably likely to imply a false or misleading representation concerning those goods … 2. in a manner that is misleading, fraudulent or deceptive in any way … "
International Trade Administration Commission of South Africa (ITAC) www.itac.org.za	To create a policy environment that promotes fair trade and to administer trade instruments (including a permitting system for import and export controls) in accordance with the: **International Trade Administration Act (Act No. 71 of 2002)**	Import controls that apply to seafood products fall under the Harmonised Commodity and Coding System (HS) tariff headings (with subheadings): • 03.02 (fresh or chilled fish excluding fillets); • 03.03 (frozen fish); • 03.04 (fish fillets); • 03.05 (dried, salted, or brined fish); • 03.06 (crustaceans); • 03.07 (mollusks); • 03.08 (other aquatic invertebrates)

[a]And amendments, where applicable.
[b]A Compulsory Specification is a technical regulation, comprising a SANS or provision of a SANS, that ensures product conformity to a specified standard. Compulsory specifications apply to certain canned, frozen, and smoked seafood products in South Africa and may specify requirements for, among others, processing facilities and persons handling and processing particular seafood products, as well as the quality, composition, packaging, and labeling of particular seafood products.
[c]Common names are designated for certain (but not all) species.

to classify commodities are the International Trade Administration Commission of South Africa (ITAC) and the South African Revenue Services (SARS) (Barendse and Francis, 2015).

Each of the aforementioned national departments enforce their respective mandates in accordance with their own Acts and Regulations, which are aligned with international guidelines and practices, such as those standards set by the Codex Alimentarius Commission (CAC, www.fao.org/fao-who-codexalimentarius). The majority of these Acts are administered and enforced at the national level, although provincial authorities and law enforcement agencies may be assigned some related responsibilities. In contrast, the DoH Acts authorize the overall enforcement of their regulations by provincial and local authorities, whereas food sampling and analysis is conducted by designated laboratories that fall under the national DoH (Chanda et al., 2010).

Challenges in Ensuring the Legality, Authenticity, and Traceability of Seafood in South Africa

Functional Fragmentation

The preceding overview of the governmental structures, legislation, and control mechanisms applicable to seafood trade in South Africa illustrates a "system" that relies on regulations, standards, enforcement, and analytical services that are scattered and administered by different national governmental departments, typical of a multiple agency food control system (FAO/WHO, 2003). Provincial and local authorities are additionally involved in enforcing seafood-related legislation drafted at the national level. Furthermore, the control of the safety and quality aspects of seafood, of processed and unprocessed products, and of import- and export-destined commodities are all segregated and thus fragmented. Although multiple agency food control systems are common worldwide, these systems present the highest probability of operational challenges when not managed effectively (Chanda et al., 2010). These challenges include a lack of coordination, duplication, and/or unclear jurisdiction of functions between the involved departments and consequent loopholes in service delivery (FAO/WHO, 2003), some of which are evident in South Africa's seafood control system. One of the most apparent challenges of fragmentation in the control of seafood trade in the country lies with the duplication of regulation and/or functions. For instance, the importation of seafood products generally involves the coordination of activities of three regulatory entities, namely the DAFF, ITAC, and SARS (Table 3.2). Similarly, both the DoH and NRCS have separate and unsynchronized regulations relating to the labeling of pre-packaged seafood products (Table 3.2), even though these requirements could be integrated and administered by a single authority.

The duplication of functions not only impedes optimal resource utilization, but also makes effective monitoring and enforcement more difficult. The scattered mandates and limited communication within and between the respective regulatory entities have led to the development and utilization of different risk management frameworks, inspection methods, and compliance verification approaches. This scenario challenges the authorities involved in the control of seafood trade to consistently manage risks across various establishment types and products, with products destined for different purposes being inspected at different frequencies and/or in different manners. The complexity of the applicable regulations also results in the industry having to fulfill multiple and different requirements for the same product (DoH/the dti/DAFF, 2013).

Legislation Loopholes

As previously mentioned, the regulations promulgated under the Acts of the DoH and NRCS make provision for product labeling (mostly for pre-packaged foodstuffs), including the requirements for a "true description" of the contents that is not misleading or confusing to the consumer, the declaration of all component ingredients in a predefined format, as well as the specification of the country of origin and/or processing. Furthermore, the recently promulgated Consumer Protection Act (under the dti, Table 3.2) protects consumers from fraudulent or deceptive labeling of goods and services and holds the entire supply chain accountable for such transgressions. Nonetheless, both the DoH and NRCS labeling regulations generally relate to pre-packaged, processed (frozen, canned, smoked) seafood products, with few or no regulations applying to fresh or whole products, regardless of whether packaged or nonpackaged. In addition, the existing labeling regulations do not extend to seafood products marketed in restaurants. These regulatory gaps provide dishonest vendors with a considerable amount of leeway to incorrectly describe or mislabel their seafood products. Moreover, whereas the regulations in place in the EU (EU 1379/2013) (Regulation (EU) No 1379/2013 of the European Parliament, 2013) specifically call for the declaration of a "commercial designation" on seafood product labels that match official lists drawn up by EU countries, no such official lists or guidance documents on the acceptable market names for seafood species are in place in South Africa, with the exception of a few scattered and largely obsolete (Barendse and Francis, 2015) common names specified for some canned fish, mollusks, and crustaceans (Industry, D. of T., 2004), frozen rock lobster (Industry, D. of T., 2003), and frozen prawns, langoustines, and crabs (Industry, D. of T., 1987). Furthermore, in spite of the requirement for the inclusion of scientific species names on import permit applications to DAFF and ITAC (Barendse and Francis, 2015), and unlike the aforementioned EU regulations (Regulation (EU) No 1379/2013 of the European Parliament, 2013), South African regulations do not specifically necessitate the inclusion of scientific names on the labels of packaged seafood products offered for sale. Given the ambiguities associated with colloquial names in the global marketplace, the mandatory requirement of scientific names on product labeling would not only promote uniformity in seafood trade, but would also assist law enforcement agencies to detect fraud or the commercialization of illegal species (Cawthorn et al., 2015). In addition, this would bring regulations in line with FAO recommendations.

With specific reference to traceability, the requirement for suppliers to ensure that fishery products are fully traceable from the fishing vessel to the end consumer is mandatory in the EU (Regulation (EC) No 178/2002 of 28 January 2002, 2002; Council Regulation, 2009; Commission Implementing Regulation, 2011), while food traceability is also among the key provisions of the Food Safety Modernization Act in the United States. However, equivalent regulations in South Africa are currently weak, fragmented, and poorly executed. As a direct consequence, traceability in local seafood supply chains is voluntary or entirely absent. With little accountability for the fish arriving at the marketplace, cases of seafood misrepresentation as previously found on the South African market are bound to occur (Cawthorn et al., 2015). The establishment of concrete legislation on traceability in South Africa is thus imperative to enhance the transparency of seafood trade, and should be placed high on the agendas of relevant policy makers.

Resource and Capacity Constraints

Financial and Human-Resource Constraints

One of the major challenges hampering effective seafood control in South Africa is the fact that national priorities differ with provincial and municipality priorities, and thus the distribution of financial and human resources to the various food control programs and regions is not uniform (DoH/the dti/DAFF, 2013). For instance, the DAFF's Directorate Compliance is tasked with ensuring compliance monitoring and enforcement of the Marine Living Resources Act (MLRA) (Act No. 18 of 1998) and related regulations, with the aid of other governmental and law enforcement agencies. Fisheries Control Officers (FCOs) appointed under this Directorate have the responsibility of, among others, inspecting fish processing establishments, restaurants, and fish shops to determine the legality of the seafood being traded (DAFF). However, surveys carried out between 2010 and 2012 by TRAFFIC East Africa in four South African provinces (Western Cape, Gauteng, KwaZulu-Natal, and Eastern Cape) revealed that the frequency of compliance inspections in restaurants and seafood outlets was low: 46% of vendors in all outlets indicated that they had never been inspected by compliance officials, and this scenario applied to 97% of the vendors from establishments in the inland province of Gauteng (Burgener et al., 2014). Illustrating the effects of this lack of compliance monitoring are the frequent opportunities for restaurants and retailers in South Africa to illegally purchase no-sale species or to purchase seafood from recreational fishers[3], and the expressed willingness of some of these operators to engage in the aforementioned

---

3. The Marine Living Resources Act (MLRA) (Act No. 18 of 1998) outlines the utilization of seafood resources for recreational fishing and legal trade in South Africa, as follows: Recreational fishers may not sell their catch (Section 20(1)). No-sale species are (1) species designated as noncommercial and thus reserved for recreational fishing, subject to the possession of a valid recreational angling permit, and daily bag limits, size limits, and closed seasons. These species may not be sold; and (2) specially protected or prohibited species that may not be captured by anyone.

practices (Burgener et al., 2014). In addition, reports of corruption and bribery of government officials suggest that this may be a key factor hindering
the effectiveness of enforcement measures (Burgener et al., 2014; Sundström,
2013; Bondaroff et al., 2015). Exacerbating the problem is the lack of special-
ized skills and improperly structured organization in critical operational areas
in the various departments (DoH/the dti/DAFF, 2013), including those relating
to seafood authentication and traceability.

Laboratory Capabilities

At present, there is a lack of accredited (ISO 17025) DNA-based food-testing
laboratories in South Africa, as well as anomaly in the distribution of such
facilities throughout the country (DoH/the dti/DAFF, 2013). Furthermore, only
a handful of official food-testing laboratories exist that are charged with ensuring compliance with applicable regulations, with their mandates largely being
fragmented based on processed versus unprocessed and import versus export
markets (Chanda et al., 2010). At present, the NRCS conducts periodic species
identification tests on selected seafood products under their jurisdiction, however, the reliability of the sensory evaluations applied for this purpose has been
questioned (Cawthorn et al., 2012a). No official laboratories currently employ
DNA-based methods for seafood authentication and origin assignment (traceability) purposes. One privately owned laboratory in South Africa offers DNA-
based seafood authentication services on a commercial basis, including DNA
barcoding. However, correspondence with this laboratory indicates that seafood
authentication testing by suppliers and retailers tends to peak around the time of
seafood mislabeling scandals, but that this is seldom maintained after the media
attention has subsided.

Initiatives aimed at stamping out seafood mislabeling in South Africa
should be based on a "proactive rather than reactive" approach, underpinned
by a structured, rigorous, and ongoing authentication program. The use of such
an approach, when reinforced by validated analytical methods, would not only
assist in identifying the links in the supply chain where mislabeling occurs, but
could provide the basis for the prosecution of illicit activities.

Positive Steps Toward Improving the Legality, Authenticity, and Traceability of Seafood in South Africa

Laying the Groundwork for Improved Seafood Authenticity Testing in South Africa

Since late 2000s, DNA sequencing has emerged as a highly informative and
reliable method for fish species identification (Puyet and Bautista, 2010). Nonetheless, the success of such authentication techniques is reliant on the availability of reference DNA sequences from a wide variety of expertly identified
species, with which unknown sequences can be compared and potentially
matched. Prior to 2009, the utilization of DNA-based methods for fish species

authentication had not been extensively explored in South Africa, and there was a dire lack of reference DNA sequence data for many domestically available species. To address this problem, a comprehensive reference library was established between 2010 and 2012, containing sufficient DNA sequence data from different mitochondrial DNA loci (16S ribosomal RNA (rRNA), 12S rRNA, and cytochrome *c* oxidase I (COI) genes, as well as the control region) to permit the unambiguous identification of 53 commonly traded fish species in South Africa (Cawthorn et al., 2012b, 2011b). In particular, this work highlighted the promise of COI barcoding for distinguishing the species origin of raw, processed, whole, or partial fish specimens (Cawthorn et al., 2011b). Even though local regulatory bodies have been slow in adopting such methodologies into their compliance testing regimes, this work laid the foundation for future applications of this kind and will undoubtedly provide an advanced level of precision to the monitoring of fish mislabeling in this country.

Seafood Awareness and Ecolabeling Programs

Although fisheries management is generally regarded as a government activity, the apparent shortcomings of these agencies in delivering sustainable fisheries (Allison, 2001; Agnew et al., 2014; Ponte, 2006) and addressing seafood traceability issues (Mariani, 2012) have led to the increasing involvement of nongovernmental organizations (NGOs) that aim to mobilize civil society to catalyze positive change. They usually seek to effect such change by altering consumer choice and/or by pressuring governments and industry, via seafood awareness campaigns and ecolabeling programs. The Marine Stewardship Council (MSC, www.msc.org) and the World Wide Fund (WWF) South Africa, through its Southern African Sustainable Seafood Initiative (WWF-SASSI, www.wwf-sassi.co.za), are two NGOs that have played leading roles in enhancing seafood market transparency in South Africa.

The Role of the Marine Stewardship Council

The MSC (www.msc.org) is the world's leading certification and ecolabeling scheme for sustainable wild-caught fisheries. MSC has received both praise and criticism (Ponte, 2006; Mariani, 2012; Gulbrandsen, 2009; Christian et al., 2013a,b; Froese and Proelss, 2012; Agnew et al., 2013; Jacquet and Pauly, 2007; Jacquet et al., 2010; Marko et al., 2011; Ward, 2008; Gutiérrez et al., 2012; Martin et al., 2012; Kaiser and Hill, 2010), but remains one of the most recognizable seafood labels, with over 19,500 products labeled in over 100 countries. MSC provides certification, based on compliance with standards for both sustainability, including target fish stock, environmental impacts and management, and chain of custody from the certified fishery to the consumer facing package. Standards are developed in consultation with scientists, the fishing industry, and conservation groups and are consistent with FAO recommendations. Notably, MSC has released data on DNA testing of their products, with plans to continue to incorporate appropriate analytical verification, such as DNA testing, to ensure authenticity of their products as part of their commitment to traceability (http://ocean-to-plate-stories.msc.org/). The Aquaculture Stewardship Council (ASC, www.asc-aqua.org) is a comparable label for aquaculture products.

After opening offices in South Africa in 2008, MSC established a strong presence in the country and the number of products bearing this ecolabel on the shelves of local retailers has continued to increase ever since. At the time of writing, a total of 73 products have approval to carry the MSC ecolabel in South Africa (Marine Stewardship Council, 2016), providing consumers with enhanced opportunities to purchase products that are certified both as sustainable and as fully traceable throughout the supply chain. While the MSC was not created specifically to deal with IUU fishing, the MSC Chain of Custody certification (CoC, Box 1) further holds promise for managing supply chains that are free of illegally caught seafood products. In particular, Woolworths (www.woolworths.co.za) currently stocks the widest range of MSC-certified products in the South African retail market (total = 32 products, frozen, chilled, and canned), and was the first retailer in the country to have their fresh seafood counters certified in accordance with the MSC CoC standard (Marine Stewardship Council, 2016). Two restaurants in Western Cape, South Africa, also are MSC approved and are therefore permitted to display the MSC ecolabel on their menu, illustrating that they source only MSC-certified products and that their suppliers have been audited against the MSC CoC standard (Marine Stewardship Council, 2016).

The South African hake trawl, targeting deep-water Cape hake (*Merluccius paradoxus*) and shallow-water Cape hake (*Merluccius capensis*), became the first fishery in Africa to gain MSC certification in 2004, and the fishery was re-certified in 2010 and 2015. With average annual catches of about 130,000 tonnes, the hake trawl contributes over 90% of the country's total commercial hake landings (trawl, longline, and handline) (SADSTIA, 2013). The hake fishery is by far the most valuable of South Africa's commercial fisheries (>50% of total value), having an annual landed value exceeding ZAR R5.2 billion (about US $330 million). More than 60% of the hake catch is currently exported, generating foreign exchange earnings of ZAR 3.5 billion (about US $220 million), with the main export markets being Spain, France, Portugal, Italy, Germany, Australia, and the United States (SADSTIA, 2013). MSC certification of the hake trawl has thus provided a proactive means to leverage increased shares in the white-fish market at the expense of competitors (e.g., Namibian, Argentinian, and Chilean hake industries, New Zealand hoki industry) (Ponte, 2006; Frohberg et al., 2006), as well as to develop new markets in parts of Europe and North America that are increasingly environmentally sensitive and virtually inaccessible to noncertified products. Moreover, the South African hake fishery contributes about 77% of the finfish caught and consumed locally (SADSTIA, 2013).

Despite the positive role that the MSC plays in promoting sustainable seafood in South Africa, the organization has come under criticism for marginalizing developing-world country fisheries, particularly small-scale ones, in their certification scheme (Ponte, 2006; Pérez-Ramírez et al., 2012; Gulbrandsen, 2009). Developing–world fisheries currently account for just 8% of the total of MSC-certified fisheries and 11% of the fisheries in full assessment (Marine Stewardship Council, 2016). Some of the traditional barriers faced by developing countries in achieving certification include open access, limited awareness about the MSC, lack of government support, poor fisheries management, data deficiencies, and ultimately the considerable associated costs. In recognition of these challenges, the MSC has established the "Developing World Program," which seeks to raise awareness of the MSC in developing countries and to provide guidance for the assessment of small-scale and data-deficient fisheries. A Sustainable Fisheries Fund has also been created to assist developing countries in going through the certification process (Marine Stewardship Council, 2016). It is further worth noting that an increasing number of fisheries are using the MSC Standard as a framework within which fishery improvements can be structured, regardless of whether they ultimately opt to become certified.

The Role of World Wide Fund Southern African Sustainable Seafood Initiative

Launched in collaboration with various networking partners under the banner of the World Wide Fund for Nature South Africa (WWF-SA) in 2004, WWF-SASSI has the core aims of fostering consumer and industry awareness around marine conservation issues, promoting voluntary compliance with applicable laws (i.e., the MLRA), and ultimately driving responsible fishing practices through a market-based approach. At the time of its inception, seafood sustainability was not a relevant priority for most suppliers and retailers in South Africa, and no retailer had taken a formal stand on ensuring that their seafood was sustainable and traceable to its origins (Kastern et al., 2014). Nonetheless, this landscape has changed considerably over the last decade, owing largely to WWF-SASSI's proactive work with consumers, retailers, suppliers, and fisheries to shift the demand and supply to more sustainable seafood options.

To aid consumers in making environmentally sound seafood choices, WWF-SASSI provides a series of tools that employ a seafood sustainability "traffic light" system that divides species into Green-list (sustainable choice), Orange-list (think twice), and Red-list (avoid) based on their stock status, the management of the relevant fishery, and the ecological impacts of the fishery. The various consumer-outreach campaigns carried out under the initiative further motivate the public to ask three key questions of their seafood: What species is it? How was it caught or farmed? Where is it from? WWF-SASSI works with the industry to reduce the impacts of destructive fishing practices and to implement an ecosystem approach to fisheries, which includes the initiation of fisheries improvement projects (FIPs) and fishery conservation projects (FCPs) among large-scale, small-scale, and subsistence fishers. To promote responsible seafood sourcing, the WWF-SASSI Retailer/Supplier Participation Scheme was developed in 2008 as a means to engage national retailers, restaurants, and their suppliers in driving transformation of source fisheries and fish farms. Nine major supply-chain stakeholders (two suppliers, four retailers, two restaurant franchises, one hotel group) have since made public, time-bound commitments to source and stock sustainable seafood, including pledges to only deal with products that are either MSC/ASC-certified, on the WWF-SASSI Green-list, or from fisheries or fish farms engaged in formal improvement programs (WWF-Southern African Sustainable Seafood Initiative). In particular, the aforementioned scheme places considerable emphasis on improving the traceability and labeling of seafood products on the local market. WWF-SASSI thus works with participants across the supply chain to identify shortcomings in traceability systems and to revise seafood product labeling to include more comprehensive information, including the declaration of the common name, scientific name, origin, and catch/production method of the species.

Standardized Seafood Naming List

With the absence of regulated market names for seafood species in South Africa, previous authors have noted that up to half of the cases of seafood misrepresentation on the local market can be linked with confounded naming practices (Cawthorn et al., 2015). For instance, conflicting common names are applied to the same species in different geographical regions in the country, generic names are provided which group various species for sale with different values, sustainability rankings, and even health concerns, imported species are often simply afforded the name of a local equivalent, while coined and fanciful names may be used to increase the market appeal of otherwise underutilized species (Barendse and Francis, 2015; Cawthorn et al., 2015). These issues are by no means unique to South Africa, and have been one of the driving forces for the compilation of recommended, acceptable, or regulated trade names for seafood species by many countries, for example, the US Food and Drug Administration (FDA) Seafood List, the Canadian Food Inspection Agency (CFIA) "Fish List," New Zealand's "Approved Fish Names," Australia's "List of Standardised Fish Names," as well as the "commercial designations" compiled by EU member states.

In line with these global trends, WWF-SA initiated the collation of such a list in consultation with a working group comprising stakeholders from government, the fishing industry, consumer bodies, suppliers, retailers, restaurateurs, research institutions, and NGOs. The SABS has subsequently agreed to advance the initiative through the compilation of SANS 1647: "Approved market names for fishery products for human consumption traded in South Africa." Currently under development, this list will aim to include the scientific names and preferably one agreed-upon market name for each species (or species group, where relevant) of local and imported fresh, frozen, and canned seafood traded in the country. While adherence to SANS are voluntary (as opposed to Compulsory Specifications) and the complexities of implementing the standard in a sector with multiple agencies and intertwined mandates remains to be tested, it is envisaged to be an important step forward in bridging the policy gap that has thus far allowed seafood mislabeling to ensue (Barendse and Francis, 2015). Adding scientific names to labels, as the EU has mandated, would be an important countermeasure.

THE CHANGING REGULATORY FRAMEWORK IN THE UNITED STATES

The United States, has recently taken steps to improve their seafood traceability system. As the second largest consumer of seafood, this has potential for global impact on the industry as it impacts all countries exporting to the United States. Under the new Food Safety Modernization Act (FSMA) requirements for both domestically produced and imported foods, the definition of food has been further clarified from that of the Food Drug and Cosmetic Act as follows, specifically listing fish and live food animals, such as molluscan shellfish, within the definition of food:

> *21 CFR §1.227 (2) Examples of food include: fruits, vegetables, fish, dairy products, eggs, raw agricultural commodities for use as food or as components of food, animal feed (including pet food), food and feed ingredients, food and feed additives, dietary supplements and dietary ingredients infant formula, beverages (including alcoholic beverages and bottled water), live food animals, bakery goods, snack food, candy and canned foods.*

Besides these changes in definition, there are also many new requirements in FSMA for processors and handlers of food products relating to traceability. The most important of these is the requirement for a Preventive controls–based food safety plan (Food Protection Plan). All facilities must have a Preventive Controls–Qualified Individual on staff to manage aspects of the Food Protection Plan. An approval process for suppliers is now required, meaning that suppliers must have traceability programs. In addition to new requirements for traceability relating to the food, processors must take condition of a processing or holding facility into account as part of hazard analysis (this will determine whether product is higher risk or not). This will include the ability to clean, zoning, and risk/control of environmental contamination. Facility condition (or ability to verify facility condition) could affect approved supplier status. By regulation, companies must use preapproved suppliers who can show traceability. In particular, the new FSMA requirements in which traceability is encompassed include economic adulteration and intentional contamination in addition to allergens, undesirable microorganisms, specific chemical hazards, and radiological hazards. There are also new requirements for environmental testing for ready-to-eat foods, and being able to tie environmental test results in to a traceability program will be a new factor for many seafood processors to consider.

Americans consumed nearly 4.6 billion pounds of seafood in 2013 (NOAA Fisheries, 2014). Due to the large rate of consumption, the domestic supply of seafood falls far short of demand in terms of quantity and desired product form. As a result, much of the seafood consumed in the United States is imported. Importing large volumes of seafood leaves the United States particularly vulnerable to challenges and risks associated with current seafood traceability schemes. In addition to economic fraud and product safety, unsustainable fishing practices are a critical concern. As a result, the United States has recently placed emphasis on combatting IUU fishing on a national scale using a risk-based strategy. A Presidential Task Force has been assembled with a focus on improving enforcement capability, catch documentation schemes, reducing complexity of the chain of custody and processing, and addressing human health risks, species misrepresentation, history of fishing violations, and mislabeling or other misrepresentation (www.nmfs.noaa.gove/ia/iuu/taskforce.html). The seafood market names currently impacted include: abalone, Atlantic and Pacific cod, blue crab, dolphin fish, grouper, red king crab, red snapper, sea cucumber, sharks, shrimps, swordfish, and tunas (albacore, bigeye, skipjack, and yellowfin).

These improvements to food traceability will heavily impact seafood regulation in the United States as well as impact countries importing into the United States. Within the United States, there are currently only a few institutions within the regulatory agencies, such as the US Customs and Border Protection (CBP), National Oceanic and Atmospheric Administration (NOAA), and the FDA, that have laboratories with full capacity to carry out DNA testing of fish if fraud is suspected. Though DNA testing is the regulatory tool of choice for seafood identification (Yancy et al., 2008; Handy et al., 2011), capacity must be increased to meet changing regulations. The changes to standardized requirements to be implemented by the United States will significantly impact the large number of nations exporting to the former. As the United States is such a large importer of seafood, changing regulations may be a step toward improved seafood traceability globally if these new traceability schemes can be implemented and validated.

CONCLUSIONS

The primary aim of traceability in food supply chains has been to regain or strengthen consumer trust by preventing or restricting the spread of adulterated or misbranded food via a system to follow the food across the entire supply chain from harvest through to the consumer (Pang et al., 2012). Additionally, traceability systems must consider and address: (1) consumer concerns over declining fish populations and the desire to purchase sustainably sourced seafood to support their core values for environmental protection and (2) consumer demand for safe, quality fish products that are truthfully represented to them. Global seafood markets and fishing industries are delicately interlinked, with the former being shaped by consumer demand and the latter determining the types of fish made available for human consumption. Illegal trade and seafood mislabeling, however, threaten to sever this link by eliminating the consumer's capacity to influence patterns of seafood exploitation through informed choices.

This chapter has detailed some of the prominent challenges faced by developed and developing countries, using the EU and South Africa as examples, in ensuring the legality, authenticity, and traceability of domestically consumed and export-bound seafood products. In the EU, continued research on implementation of DNA-based traceability methods is needed. In addition, though the EU has a fully developed regulatory framework, implementation by member states may differ in efficacy and complexity. The United States has new laws mandating greater traceability, but has limited resources to implement the regulatory structure required using the current resource base and is moving toward a system of high fees and civil and criminal penalties to increase the resource base. Regardless, control of imports will prove to be difficult, and domestic producers will bear the brunt of regulatory enforcement actions. South Africa is challenged by weak governance and inadequate law enforcement. In the EU, harmonization of regulations and implementation of new tools must be done across many member states, and may therefore suffer from fragmentation in management of issues, much like South Africa. Nonetheless, we have also highlighted how positive change can emerge at the interface between science, the media, and industry, and how NGOs can play a crucial role in drawing together disparate actors with different knowledge to jointly drive the improvement of seafood market transparency in both developed and developing nations. This continued collaboration and progression toward improved seafood management and regulation is critical to the sustainability of the industry and exerts impact on the development of improved or new regulations.

Overall, the seafood industry would benefit from global harmonization in traceability requirements, including the use of DNA testing. Countries or jurisdictions with established regulations may find implementation of paradigm shifts, such as the FSMA, to be slow and difficult, as fitting new rules into an existing framework involves interplay between policy makers, regulators, and industry. Countries just developing their regulations are faced with the difficulty of putting a new system into place, enforcing new rules, and educating regulators. Concerted and collaborative international efforts may serve to ease both these transitions. Increased input from international groups, such as FAO, may help align existing regulations and form a standardized set of criteria. Citizens also play a role in encouraging improved traceability, and working with NGOs and the scientific community can continue to drive needed changes in seafood supply chain traceability and labeling practices.

REFERENCES

Agnew, D.J., et al., 2009. Estimating the worldwide extent of illegal fishing. PLoS One 4, e4570. http://dx.doi.org/10.1371/journal.pone.0004570.

Agnew, D.J., et al., 2013. Rebuttal to Froese and Proelss 'evaluation and legal assessment of certified seafood'. Marine Policy 38, 551–553.

Agnew, D.J., Gutierrez, N.L., Stern-Pirlot, A., Hoggarth, D.D., 2014. The MSC experience: developing an operational certification standard and a market incentive to improve fishery sustainability. ICES Journal of Marine Science 71, 216–225.

Allison, E.H., 2001. Big laws, small catches: global ocean governance and the fisheries crisis. Journal of International Development 13, 933–950.

Barendse, J., Francis, J., 2015. Towards a standard nomenclature for seafood species to promote more sustainable seafood trade in South Africa. Marine Policy 53, 180–187.

Blancou, J., 2001. A History of the Traceability of Animals and Animal Products.

Bondaroff, T.N.P., Reitano, T., van der Werf, W., 2015. The Illegal Fishing and Organized Crime Nexus: Illegal Fishing as Transnational Organized Crime. The Global Initiative Against Transnational Organized Crime/The Black Fish, Geneva.

Burgener, M., Kastern, C., Duncan, J., Barendse, J., McLean, B., Cawthorn, D.M., Okes, N., 2014. From Boat to Plate: Linking the Seafood Consumer and Supply Chain. . http://dx.doi.org/10.13140/RG.2.1.2814.0967.

Cawthorn, D.M., Steinman, H.A., Witthuhn, R.C., 2011a. Evaluating the availability of fish species on the South African market and the factors undermining sustainability and consumer choice. Food Control 22, 1748–1759.

Cawthorn, D.-M., Steinman, H.A., Witthuhn, R.C., 2011b. Establishment of a mitochondrial DNA sequence database for identification of fish species commercially available in South Africa. Molecular Ecology Resources 11, 979–991.

Cawthorn, D.M., Steinman, H.A., Witthuhn, R.C., 2012a. DNA barcoding reveals a high incidence of fish species misrepresentation and substitution on the South African market. Food Research International 46, 30–40.

Cawthorn, D.M., Steinman, H.A., Witthuhn, R.C., 2012b. Evaluation of the 16S and 12S rRNA genes as universal markers for the identification of commercial fish species in South Africa. Gene 491, 40–48.

Cawthorn, D.M., Duncan, J., Kastern, C., Francis, J., Hoffman, L.C., 2015. Fish species substitution and misnaming in South Africa: an economic, safety and sustainability conundrum revisited. Food Chemistry 185, 165–181.

Chanda, R.R., Fincham, R.J., Venter, P., 2010. A review of the South African food control system: challenges of fragmentation. Food Control 21, 816–824.

Charlesbois, S., Sterling, B., Haratifar, S., Kyaw Naing, S., 2014. Comparison of global food traceability regulations and requirements. Comprehensive Reviews of Food Science and Food Safety 13 (5), 1104–1123.

Christian, C., et al., 2013a. A review of formal objections to Marine Stewardship Council fisheries certifications. Biological Conservation 161, 10–17.

Christian, C., et al., 2013b. Not 'the best environmental choice in seafood': a response to Gutiérrez and Agnew (2013). Biological Conservation 165, 214–215.

Commission Implementing Regulation (EU) No 404/2011 of 8 April 2011 laying down detailed rules for the implementation of C.R. (EC) 1224/2009 establishing a Community control system for ensuring compliance with the rules of the Common Fisheries Policy. Official Journal of the European Union L112, 2011, 1–153.

Council Regulation (EC) No 1224/2009 of 20 November 2009 establishing a Community control system for ensuring compliance with the rules of the common fisheries policy. Official Journal of the European Union L343, 2009, 1–50.

Council, 2003. Council Regulation (EC) N O 21/2004 Establishing a System for the Identification and Registration of Ovine and Caprine Animals and Amending Regulation (EC) No 1782/2003 and Directives 92/102/EEC and 64/432/EEC, as Amended.

Council, 2009. Regulation (EC) No 1224/2009 Establishing a Community Control System for Ensuring Compliance With the Rules of the Common Fisheries Policy, Amending Regulations (EC) No 847/96, (EC) No 2371/2002, (EC) No 811/2004, (EC) No 768/2005, (EC) No 2115/2005, (EC) No 2166/2005, (EC) No 388/2006, (EC) No 509/2007, (EC) No 676/2007, (EC) No 1098/2007, (EC) No 1300/2008, (EC) No 1342/2008 and Repealing Regulations (EEC) No 2847/93, (EC) No 1627/94 and (EC) No 1966/2006 as Amended.

Delgado, C.L., Wada, N., Rosegrant, M.W., Siet, M., Ahmed, M., 2003. Outlook for Fish to 2020: Meeting Global Demand. IFPRI/World Fish Center, Washington, DC/Malaysia.

Department of Agriculture, Forestry and Fisheries, 2014. Status of the South African Marine Fishery Resources. DAFF, Cape Town.

Department of Agriculture, Forestry and Fisheries (DAFF). Fisheries Management: Monitoring, Control and Surveillance, Directorate Compliance. Available at: http://www.daff.gov.za/daff-web3/Branches/Fisheries-Management/Monitoring-Control-and-Surveillance/compliances.

Department of Health (DoH)/Department of Trade and Industry (the dti)/Department of Agriculture, Forestry and Fisheries (DAFF), 2013. Food Safety and Food Control in South Africa: Specific Reference to Meat Labelling. DAFF/DoH, dti, Pretoria.

Domdouzis, K., Kumar, B., Anumba, C., 2007. Radio-frequency identification (RFID) applications: a brief introduction. Advanced Engineering Informatics 21, 350–355.

European Commission, 2001. Regulation (EC) No 2065/2001 Laying Down Detailed Rules for the Application of Council Regulation (EC) No 104/2000 as Regards Informing Consumers About Fishery and Aquaculture Products, as Amended.

European Commission, Directorate-General Press and Communication & Office for Official Publications of the European Communities, 2004a. From Farm to Fork: Safe Food for Europe's Consumers. Office for Official Publications of the European Communities.

European Commission, 2004b. Regulation (EC) No 641/2004 on Detailed Rules for the Implementation of Regulation (EC) No 1829/2003 of the European Parliament and of the Council as Regards the Application for the Authorisation of New Genetically Modified Food and Feed, the Notification of Existing Products and Adventitious or Technically Unavoidable Presence of Genetically Modified Material Which Has Benefited from a Favourable Risk Evaluation.

European Commission, 2008. Regulation (EC) No 504/2008 Implementing Council Directives 90/426/EEC and 90/427/EEC as Regards Methods for the Identification of Equidae, as Amended.

European Commission, 2011a. Commission Regulation (EU) No 619/2011 Laying Down the Methods of Sampling and Analysis for the Official Control of Feed as Regards Presence of Genetically Modified Material for Which an Authroisation Procedure Is Pending or the Authorisaton of Which Has Expired.

European Commission, 2011b. Implementing Regulation (EU) No. 404/2011 Laying Down Detailed Rules for the Implementation of Council Regulation (EC) No. 1224/2009 Establishing a Community Control System for Ensuring Compliance With the Rules of the Common Fisheries Policy.

European Commission and Directorate-General for Agriculture and Rural Development, 2012. The Common Agricultural Policy: A Story to Be Continued. EUR-OP.

European Commission, 2013. Implementing Regulation (EU) No 503/2013 on Applications for Authorisation of Genetically Modified Food and Feed in Accordance with Regulation (EC) No 1829/2003 of the European Parliament and of the Council and Amending Commission Regulations (EC) No 641/2004 and (EC) No 1981/2006 Text With EEA Relevance.

European Council, 2008a. Regulation (EC) No 1005/2008 Establishing a Community System to Prevent, Deter and Eliminate Illegal, Unreported and Unregulated Fishing, Amending Regulations (EEC) No 2847/93, (EC) No 1936/2001 and (EC) No 601/2004 and Repealing Regulations (EC) No 1093/94 and (EC) No 1447/1999, as Amended.

European Council, 2008b. Regulation (EC) No 1006/2008 Concerning Authorisations for Fishing Activities of Community Fishing Vessels outside Community Waters and the Access of Third Country Vessels to Community Waters, Amending Regulations (EEC) No 2847/93 and (EC) No 1627/94 and Repealing Regulation (EC) No 3317/94.

European Parliament and Council, 2001. Directive 2001/18/EC of the European Parliament and of the Council on the Deliberate Release into the Environment of Genetically Modified Organisms and Repealing Council Directive 90/222/EEC.

European Parliament and Council, 2003a. Regulation (EC) No 1829/2003 of the European Parliament and of the Council on Genetically Modified Food and Feed, as Amended.

European Parliament and Council, 2003b. Regulation (EC) No 1830/2003 of the European Parliament and of the Council Concerning the Traceability and Labelling of Genetically Modified Organisms and the Traceability of Food and Feed Products Produced from Genetically Modified Organisms and Amending Directive 2001/18/EC.

European Parliament and Council, 2013. Regulation (EU) No 1380/2013 on the Common Fisheries Policy, Amending Council Regulations (EC) No 1954/2003 and (EC) No 1224/2009 and Repealing Council Regulations (EC) No 2371/2002 and (EC) No 639/2004 and Council Decision 2004/585/EC.

European Parliament and the Council, 1997. Regulation (EC) No 258/97 of the European Parliament and of the Council Concerning Novel Food and Novel Food Ingredients, as Amended.

European Parliament and the Council, 2002. Regulation (EC) No 178/2002 Laying Down the General Principles and Requirements of Food Law, Establishing the European Food Safety Authority and Laying Down Procedures in Matters of Food Safety, as Amended.

European Parliament and the Council, 2011. Regulation (EU) No 1169/2011 on the Provision of Food Information to Consumers, Amending Regulations (EC) No 1924/2006 and (EC) No 1925/2006 of the European Parliament and of the Council, and Repealing Commission Directive 87/250/EEC, Council Directive 90/496/EEC, Commission Directive 1999/10/EC, Directive 2000/13/EC of the European Parliament and of the Council, Commission Directives 2002/67/EC and 2008/5/EC and Commission Regulation (EC) No 608/2004.

FAO, 1995. Code of Conduct for Responsible Fisheries.

Food and Agriculture Organization (FAO), 2014a. FAO Yearbook: Fishery and Aquaculture Statistics 2012. FAO, Rome.

Food and Agriculture Organization (FAO), 2014b. The State of World Fisheries and Aquaculture 2014. FAO, Rome.

Food and Agriculture Organization/World Health Organization, 2003. Assuring Food Safety and Quality: Guidelines for Strengthening National Food Control Systems. FAO/WHO, Rome/Geneva. Available at: ftp://ftp.fao.org/docrep/fao/006/y8705e/y8705e00.pdf.

Froese, R., Proelss, A., 2012. Evaluation and legal assessment of certified seafood. Marine Policy 36, 1284–1289.

Frohberg, K., Grote, U., Winter, E., 2006. EU food safety standards, traceability and other regulations. In: Invited Paper Prepared for Presentation at the International Association of Agricultural Economists Conference, Gold Coast, Australia, August 12–18, 2006.

Glover, K.A., Haug, T., Øien, N., Walløe, L., Lindblom, L., Seliussen, B.B., Skaug, H.J., 2012. The Norwegian minke whale DNA register: a data base monitoring commercial harvest and trade of whale products. Fish and Fisheries 13, 313–332. http://dx.doi.org/10.1111/j.1467-2979.2011.00447.x.

Griffiths, C.L., Robinson, T.B., Lange, L., Mead, A., 2010. Marine biodiversity in South Africa: an evaluation of current states of knowledge. PLoS One 5, e12008. http://dx.doi.org/10.1371/journal.pone.0012008.

Gulbrandsen, L.H., 2009. The emergence and effectiveness of the Marine Stewardship Council. Marine Policy 33, 654–660.

Gutiérrez, N.L., et al., 2012. Eco-label conveys reliable information on fish stock health to seafood consumers. PLoS One 7, 1–8.

Handy, S.M., Deeds, J.R., Ivanova, N.V., Hebert, P.D.N., Hanner, R.H., Ormos, A., Weigt, L.A., Moore, M.M., Yancy, H.F., 2011. A single-laboratory validated method for the generation of DNA barcodes for the identification of fish for regulatory compliance. Journal of AOAC International 94 (1), 201–210.

Hauck, M., Kroese, M., 2006. Fisheries compliance in South Africa: a decade of challenges and reform 1994–2004. Marine Policy 30, 74–83.

Hsu, Yu-C., Chen, An-P., Wang, C.-H., 2008. A RFID-enabled traceability system for the supply chain of live fish. IEEE 81–86. http://dx.doi.org/10.1109/ICAL.2008.4636124.

Industry, D. of T., 1987. Compulsory specification for frozen shrimps (prawns), langoustines and crabs (VC8031/1987). Government Gazette 10614, 1–73.

Industry, D. of T, 2003. Compulsory specification for frozen rock lobster and frozen lobster products derived therefrom (VC8020/2003). Government Gazette 25171, 1–46.

Industry, D. of T, 2004. Compulsory specification for the manufacture, production, processing and treatment of canned fish, canned marine molluscs and canned crustaceans. VC8014/2004. Government Gazette 26530, 1–81.

Jacquet, J., et al., 2010. Seafood stewardship in crisis. Nature 467, 28–29.

Jacquet, J.L., Pauly, D., 2007. The rise of seafood awareness campaigns in an era of collapsing fisheries. Marine Policy 31, 308–313.

Kaiser, M.J., Hill, L., 2010. Marine stewardship: a force for good. Nature 467, 531.

Kastern, C., Rainer, S., Duncan, J., 2014. WWF-SASSI Retailer/Supplier Participation Scheme. WWF-South Africa, Cape Town.

Lynch, D., Vogel, D., 2001. The Regulation of GMOs in Europe and the United States: A Case-Study of Contemporary European Regulatory Politics. Available at: http://www.cfr.org/agricultural-policy/regulation-gmos-europe-united-states-case-study-contemporary-european-regulatory-politics/p8688.

Mariani, S., 2012. Seafood genetic identification: aiming our pipettes at the right targets. Frontiers in Ecology and the Environment 10, 10.

Marine Stewardship Council, 2016. Available at: https://www.msc.org/.

Marko, P.B., Nance, H.A., Guynn, K.D., 2011. Genetic detection of mislabeled fish from a certified sustainable fishery. Current Biology 21, R621–R622.

Martin, S.M., Cambridge, T., Grieve, C., Nimmo, F.M., Agnew, D.J., 2012. An evaluation of environmental changes within fisheries involved in the Marine Stewardship Council certification scheme. Reviews in Fisheries Science 20, 61–69.

Miraglia, M., et al., 2004. Detection and traceability of genetically modified organisms in the food production chain. Food and Chemical Toxicology 42, 1157–1180.

National Fisheries Institute, 2015. CTE and KDE for Seafood c/o Barbara Blakistone. US Seafood Traceability Implementation Guide, Mc Lean, VA.

New Analytical Approaches for Verifying the Origin of Food, 2013. Woodhead Publishing Limited.

Nielsen, E.E., et al., 2012. Gene-Associated Markers Provide Tools for Tackling Illegal Fishing and False Eco-certification. Nature Communications 3, 851.

NOAA Fisheries, 2014. Fisheries of the United States, 2013, A Statistical Snapshot of 2013 Fish Landings. Available: https://www.st.nmfs.noaa.gov/Assets/commercial/fus/fus13/materials/FUS2013_FactSheet_FINAL.pdf.

Ogden, R., Linacre, A., 2015. Wildlife forensic science: a review of genetic geographic origin assignment. Forensic Science International: Genetics. http://dx.doi.org/10.1016/j.fsigen.2015.02.008.

Olsen, P., Borit, M., 2013. How to define traceability. Trends in Food Science and Technology 29, 142–150.

Organisation for Economic Co-operation and Development (OECD), 2010. Globalisation in Fisheries and Aquaculture: Opportunities and Challenges. OECD Publishing, Paris.

Pang, Z., Chen, Q., Han, W., Zheng, L., 2012. Value-centric design of the internet-of-things solution for food supply chain: value creation, sensor portfolio and information fusion. Information Systems Frontiers 17.

Pauly, D., 2009. Beyond duplicity and ignorance in global fisheries. Scientia Marina 73, 215–224.

Pérez-Ramírez, M., Phillips, B., Lluch-Belda, D., Lluch-Cota, S., 2012. Perspectives for implementing fisheries certification in developing countries. Marine Policy 36, 297–302.

Ponte, S., 2006. Ecolabels and Fish Trade: Marine Stewardship Council Certification and the South African Hake Industry. Tralac Working Paper 9. Trade Law Centre for Southern Africa, Stellenbosch.

Pramod, G., Nakamura, K., Pitcher, T.J., Delagran, L., 2014. Estimates of illegal and unreported fish in seafood imports to the USA. Marine Policy 48, 102–113.

Puyet, A., Bautista, J.M., 2010. Detection of adulterations: identification of seafood species. In: Handbook of Seafood and Seafood Products Analysis. CRC Press, Florida.

Regulation (EU) No 1379/2013 of the European Parliament and of the Council of 11 December on the common organisation of the markets in fishery and aquaculture products, amending C.R. (EC) 1184/2006 and (EC) 1224/2009 and repealing C.R. (EC) 104/2000. Official Journal of the European Union L354, 2013, 1–21.

Regulation (EC) No 178/2002 of 28 January 2002 laying down the general principles and requirements of food law, establishing the European Food Safety Authority and laying down procedures in matters of food safety. Official Journal of the European Communities L31, 2002, 1–24.

Regulation (EU) No 1379/2013 on the Common Organisation of the Markets in Fishery and Aquaculture Products, Amending Council Regulations (EC) No 1184/2006 and (EC) No 1224/2009 and Repealing Council Regulation (EC) No 104/2000, 2013. .

Seimenis, A.M., Economides, P.A., 2002. The Role of Veterinary Services in the Food Chain 'From the Stable to the Table'. OIE, pp. 307–319.

Sengstschmid, H., Sprong, N., Schmid, O., Stockebrand, N., Stolz, H., Spiller, A., 2011. EU Ecolabel for Food and Feed Products – Feasibility Study (ENV.C.1/ETU/2010/0025). Commissioned by DG Environment, European Commission.

Smith, J.L.B., Smith, M.M., 1966. Fishes of the Tsitsikamma Coastal National Park. National Parks Board, Pretoria.

Sundström, A., 2013. Corruption in the commons: why bribery hampers enforcement of environmental regulations in South African fisheries. International Journal of the Commons 7, 454–472.

The South African Deep Sea Trawling Industry Association (SADSTIA), 2013. Available at: http://www.sadstia.co.za/.

United Nations Convention on the Law of the Sea (UNCLOS), 1982. 1833 UNTS 3; 21 ILM 1261 (Adopted 10 December 1982, Entered into Force 16 November 1994). Available at: http://www. un.org/Depts/los/convention_agreements/convention_overview_convention.htm.

Von Der Heyden, S., Barendse, J., Seebregts, A.J., Matthee, C.A., 2010. Misleading the masses: detection of mislabelled and substituted frozen fish products in South Africa. ICES Journal of Marine Science 67, 176–185.

Ward, T., 2008. Barriers to biodiversity conservation in marine fishery certification. Fish and Fisheries 169–177.

WWF-Southern African Sustainable Seafood Initiative. Available at: http://wwfsassi.co.za/.

Yancy, H.F., Zemlark, T.S., Mason, J.A., Washington, J.D., Tenge, B.J., Nguyen, N.T., Barnett, J.D., Savary, W.E., Hill, W.E., Moore, M.M., Fry, F.S., Randolph, S.C., Rogers, P.L., Hebert, P.D.N., 2008. Potential use of DNA barcodes in regulatory science: applications of the Regulatory Fish Encyclopedia. Journal of Food Protection 71, 210–217.

Chapter 4

IUU Fishing and Impact on the Seafood Industry

Dana D. Miller, U. Rashid Sumaila
University of British Columbia, Vancouver, BC, Canada

Species substitution and the mislabeling of seafood products potentially have numerous repercussions for consumers, the global fishing industry and the environment. As detailed in Chapter 1, consequences borne by society may include human health risks (e.g., Cohen et al., 2009; Ling et al., 2009; Triantafyllidis et al., 2010) and economic losses (e.g., Miller and Mariani, 2010; Garcia-Vazquez et al., 2011; Wong and Hanner, 2008). Depending on the jurisdiction and circumstances under which it occurs, seafood fraud can also be considered an unlawful and criminal act (Miller and Mariani, 2010; Miller et al., 2012b).

Understanding the underlying reasons for why seafood fraud occurs may be just as important as understanding how it occurs and the direct, or indirect implications it may have for individual consumers or society. Prior to the marketing, labeling, packaging and processing of seafood, it needs to be sourced and harvested either from aquaculture farms or from wild fisheries. In regards to wild fish that are caught and then subsequently mislabeled, the capture of this fish may also be unlawful in some circumstances. Mislabeling may then be used as a strategy for concealing illegally caught fish (Miller et al., 2012b). If effective measures are not taken to combat seafood fraud, then this strategy may remain an attractive technique for laundering illegally or unsustainably caught fish. Thus, seafood mislabeling may facilitate the continuance of unsustainable and irresponsible fishing practices, or what has commonly been referred to as illegal, unreported and unregulated (IUU) fishing (FAO, 2001; Theilen, 2013).

ILLEGAL, UNREPORTED AND UNREGULATED FISHING DEFINED

Rather than simply referring to "illegal fishing" as a problem facing global fisheries, the term "IUU fishing" is frequently used instead. This term has emerged as it has been recognized that depending on the type of fishing activity, who is fishing and where the fishing takes place, an act of fishing may or may not technically or strictly qualify as "illegal" activity. However, fishing activities that

Seafood Authenticity and Traceability. http://dx.doi.org/10.1016/B978-0-12-801592-6.00004-8
83

are "unreported" and/or "unregulated" may also be considered threats to marine habitats and the sustainability of managed fisheries (Bray, 2001).

IUU fishing may be deliberate or unintentional, and occurs both within the exclusive economic zones (EEZs) of coastal States and in the high seas (Bray, 2001). The high seas encompass waters that are in marine areas beyond national jurisdiction, transcending the outer boundaries of EEZs, 200 nautical miles out from the shoreline of any coastal State (UN, 1982). When a vessel enters the high seas, the responsibility for monitoring and controlling the activities of the vessel shifts from the coastal State to the state in which the vessel has been registered under, commonly known as a vessel's "flag State." Regional fisheries management organizations (RFMOs) have been formed by countries that have fishing interests in shared areas, typically located either exclusively or partially within high seas areas (UN, 1995). RFMOs have a responsibility to facilitate intergovernmental cooperation in fisheries management and often deal with issues relating to IUU fishing. However, in part because of the regulatory challenges that are currently associated with fishing activities in the high seas, closing this area to fishing is an idea that has been proposed amongst academics, and is currently an issue of debate in international fora (White and Costello, 2014; Sumaila et al., 2015).

The first formal mention of the term "IUU fishing" was in 1997, during a meeting of the Commission for the Conservation of Antarctic Marine Living Resources (CCAMLR) (Doulman, 2000). CCAMLR is an RFMO that was established in 1982 with an objective of conserving Antarctic marine life. CCAMLR routinely agrees upon conservation measures that determine the use of marine living resources in CCAMLR's "Convention Area" and these measures are binding on all Contracting Parties to the CCAMLR Convention (CCAMLR, 1980). CCAMLR's Convention Area is an area south of the Antarctic Convergence, covering about 10% of the Earth's surface. IUU fishing evolved as a term from discussions concerning illegal and/or non-CCAMLR-compliant fishing activities by Parties to the Convention (illegal and unreported) and non-Parties to the Convention (illegal and unregulated) (Doulman, 2000). Since the relatively recent emergence of the term IUU fishing, it is now used frequently not only by CCAMLR, but also within academic literature and by other large international organizations including the UN Food and Agriculture Organization (FAO).

In recognition of the growing importance of IUU fishing as a problem facing the world's oceans, the UN developed an International Plan of Action to Prevent, Deter and Eliminate IUU Fishing (IPOA-IUU), within the framework of the Code of Conduct for responsible fisheries, as a voluntary instrument. The FAO Committee on Fisheries (COFI) adopted the IPOA-IUU by consensus in 2001 and within this non-binding instrument, formal definitions for IUU fishing have been given (FAO, 2001, Articles 3.1–3.3):

3.1 Illegal fishing refers to fishing activities:

    3.1.1 conducted by national or foreign vessels in waters under the jurisdiction of a State, without the permission of that State, or in contravention of its laws and regulations;

3.1.2 conducted by vessels flying the flag of States that are parties to a relevant RFMO but operate in contravention of the conservation and management measures adopted by that organization and by which the States are bound, or relevant provisions of the applicable international law; or

3.1.3 in violation of national laws or international obligations, including those undertaken by cooperating States to a relevant RFMO.

3.2 Unreported fishing refers to fishing activities:

    3.2.1 which have not been reported, or have been misreported, to the relevant national authority, in contravention of national laws and regulations; or

    3.2.2 undertaken in the area of competence of a relevant RFMO which have not been reported or have been misreported, in contravention of the reporting procedures of that organization.

3.3 Unregulated fishing refers to fishing activities:

    3.3.1 in the area of application of a relevant RFMO that are conducted by vessels without nationality, or by those flying the flag of a State not party to that organization, or by a fishing entity, in a manner that is not consistent with or contravenes the conservation and management measures of that organization; or

    3.3.2 in areas or for fish stocks in relation to which there are no applicable conservation or management measures and where such fishing activities are conducted in a manner inconsistent with State responsibilities for the conservation of living marine resources under international law.

The above definitions have also been referred to within the FAO Agreement on Port State Measures to Prevent, Deter and Eliminate Illegal, Unreported and Unregulated Fishing (PSMA) (FAO, 2009a). Adopted in 2009, the PSMA has been in force since June of 2016, when it became legally binding for the 29 countries and a regional organization that have officially adhered to it.

Despite the existence of an official definition for IUU fishing, this term is still in its infancy, and is therefore often used in an inconsistent manner or equated to just "illegal fishing." It has also been argued that under the IPOA-IUU definition, both "unreported" fishing and "unregulated" fishing are described as subcategories for illegal fishing and thus "IUU fishing" actually does simply refer to just "illegal fishing" (Theilen, 2013). Illegal or legal fishing that is "unreported" and/or "unregulated" can be uniquely problematic though. Both forms of fishing activities are either completely, or inaccurately unaccounted for when the status of fisheries stocks are evaluated, which may lead to the implementation of unsustainable stock management plans. In the absence of monitoring and regulation, it is more difficult to evaluate the impact that fishing activities may be having on a particular stock (Doulman, 2000). To develop an operational strategy for tackling "illegal" fishing activity however, a more helpful differentiation of subcategories would perhaps be to focus on the difference between violation of national law and of international obligations, as referred to within the IPOA-IUU definition (Theilen, 2013). Regardless of official interpretations, it should be recognized that the mainstream use and meaning of the term "IUU fishing" is often complicated, reflecting the similarly complicated landscape of international fisheries governance.

EXAMPLES OF ILLEGAL, UNREPORTED AND UNREGULATED FISHING

IUU fishing exists in many different forms; some imposing greater consequences on the environment, law-abiding fishers and to society than others. Examples of IUU fishing include but are not limited to: catching over a permitted amount; catching and retaining non-permitted species or species over or under a permitted size limit; the utilization of banned gear; fishing in a closed area or an area where a vessel is not permitted because of its nationality or size; fishing during a closed season; and unreported or unregulated fishing as described previously in the IUU Fishing Defined section (Bray, 2001). IUU fishing may be carried out by small- or large-scale fishing vessels, be deliberate or unintentional, and may exist as infrequent isolated events or as a typical mode of operation. Individual vessels may partake in IUU fishing alone, or they may be part of a larger syndicate with multiple layers of corporate ownership disguised through registration in numerous countries (Griggs and Lugten, 2007). It is difficult to say for certain which forms of IUU fishing are the most damaging from a global, collective point of view, in part because it has proven challenging to quantify the global extent of IUU fishing (Agnew et al., 2009). Regardless, perhaps the forms that are the most disconcerting are those that are the most deliberate and the most effective in generating profit while facilitating the avoidance of retribution.

Fishing vessels that deliberately and frequently engage in IUU fishing activity often employ tactics to avoid detection or to circumvent prosecution (Flothmann et al., 2010). One frustratingly effective tactic is the use of "flags of convenience" (FoCs) or "flags of non-compliance" (FoNCs). As mentioned previously, fishing activities that take place within 200 nautical miles from the shoreline of any coastal State (the EEZ of that State) fall under the jurisdiction and control of the adjacent coastal State. Outside these waters, in the high seas, fishing activities fall under the jurisdiction of the State that a vessel is registered under, also known as a vessel's flag State. Under international law, a genuine link should technically exist between a vessel and its flag State (UN, 1982). However, the practice of registering vessels under foreign flags, i.e., FoCs is fairly common within the global maritime sector, including amongst some large fishing vessels and vessels that transport fish (Doulman, 2000; Miller and Sumaila, 2014). The use of FoCs often allows vessel owners to cut labor costs, avoid taxation and reduce expenses relating to other legal requirements. The disconnection between the nationalities of vessels and vessel owners in these situations may also limit the ability of flag States to monitor and effectively enforce legislation upon all vessels that fly their flag. Furthermore, there are many States that have not signed, ratified or acceded to international agreements developed to introduce minimum standards relating to crew health and safety, marine pollution, the conservation of living natural resources, etc. There also exist States that may officially be legally bound to these agreements but for whatever reason are not able to, or refuse to implement and/or enforce treaty provisions. When a vessel owner chooses to use a flag that is from a State that exhibits a consistent pattern of failure in compliance with its international obligations, this practice has been referred to as using an FoNC (FAO, 2009b). FoNCs have been found to be more common amongst vessels that have been implicated in IUU fishing activity compared to all other large-scale fishing vessels (Miller and Sumaila, 2014). Thus, it is likely that some IUU fishing vessel owners preferentially register their vessels under FoNCs to escape the monitoring, control and surveillance (MCS) activities of more environmentally responsible flag States. Although a formal list of FoNCs does not exist, the EU maintains a list of "non-cooperating third countries" that have been identified for their poor track record in cooperating with international efforts to combat IUU fishing (current list available at http://ec.europa.eu/fisheries/cfp/illegal_fishing/info/index_en.htm). Under the EU Regulation to prevent, deter and eliminate IUU fishing (the EU-IUU Regulation), sanctioning mechanisms exist to block access at EU ports to EU markets and services to vessels registered under the flags of non-cooperating third countries (EC, 2008). IUU vessels have also been known to frequently change their flag, a practice commonly known as "flag-hopping," which may be done to avoid prosecution under the fishing regulations relevant to a particular fishing area or flag State (Birnie, 1993; Miller and Sumaila, 2014).

The frustration associated with the use of the tactics described previously (FoCs, FoNCs and flag-hopping) is that often they characterize "unregulated" fishing, which may not necessarily also be "illegal" and/or punishable by law. If a vessel is fishing within an RFMO area in a manner that is inconsistent with the fishing rules agreed to by that RFMO and the vessel is flying the flag of a State that is not a Member of the RFMO, the vessel may not technically be strictly bound to RFMO rules (Bray, 2001). Although the activities of these fishers may be undeniably unsustainable and damaging to the environment, it can be difficult to halt the operation of these vessels. IUU vessels may also operate in the waters that straddle the boundary between EEZs and the high seas, ducking in and out of areas where they are not permitted to fish in. In these situations, it can also be extremely difficult to apprehend and punish offenders, particularly if the coastal State does not have the resources to effectively monitor these waters and to pursue vessels that are observed illegal fishing (Bray, 2001).

Vessels may try to avoid detection while engaging in IUU activity through switching off required vessel tracking devices (Windward, 2014) or again, by taking advantage of weaknesses in international legislation that limits transparency within the industry. Deficiencies in the use of International Maritime Organization (IMO) numbers and Automatic Identification System (AIS) vessel tracking are a few examples of such weaknesses. An IMO number is a unique permanent identification number that is designed to ensure that a vessel can be identified and linked to any existing historical records throughout its lifetime despite changes in ownership, flag or name. The AIS tracking system was designed for navigational purposes and it automatically transmits information on vessel identity and navigational status to other ships- and shore-based facilities. The primary purpose of AIS is to assist in the prevention of navigational accidents, although if use is mandated and enforced, it could also be used to supplement monitoring and surveillance of fishing vessel activity. Both IMO numbers and AIS are required under the International Convention for the Safety of Life at Sea (SOLAS) for passenger vessels and cargo ships 300 GT or larger operating under the flags of Member States party to the Convention (IMO, 1974). For fishing vessels, there is no requirement to have an IMO number and registration is only voluntary. It is up to the flag State to determine whether or not the requirement for AIS is applicable. Because fishing vessels are not required under international law to have an IMO number, or actively use AIS vessel tracking, depending on where they are operating and under what flag they are registered the detection and identification of IUU vessels that take advantage of these exemptions can be difficult.

In order for IUU fishing activities to be profitable, once fish has been caught, it needs to be brought to shore and sold into the global seafood market. In recent years, access to seafood markets has become increasingly regulated, particularly in Europe and in North America (e.g., EC, 2008). Vessels that have been officially recognized for their involvement in IUU fishing activities may have difficulties landing their catches at certain ports or in certain countries. In addition, and as mentioned previously, vessels that are flagged by States that have been deemed non-cooperative in efforts to stem IUU fishing by the European Commission (EC) may not be allowed to sell fish to any EU Member States (EC, 2008). To circumvent these restrictions, and also to save on fuel costs by not having to return to port to unload their catch, IUU vessels may transship their catch onto a fish-carrying vessel at sea. This vessel is usually a refrigerated cargo vessel, also known as a reefer. Reefers can load up with catches from numerous different fishing vessels while at sea before returning to port to offload and sell the combined catch into the seafood market. Depending on where fishing activities occur, transshipments at sea may not be allowed or may need to be reported and/or monitored (EJF, 2013). Again, depending on where fishing activities occur and also what market the catch is destined for, upon transshipment, the transferred catch from each fishing vessel may need to be clearly labeled with information on the catching vessel, the catch area and the species of catch. Despite these requirements, the process of transshipping at sea provides an opportunity for the intentional mislabeling of fish and laundering of IUU catch amongst legitimate catch.

COMBATTING ILLEGAL, UNREPORTED AND UNREGULATED FISHING

Just as IUU fishing has taken on many different forms, many different strategies have been developed to combat IUU fishing by national governments and through international, collaborative efforts. On a local scale, coastal States govern fishing activities within their waters through their respective MCS regimes (Bray, 2001). The enforcement of these regimes may entail the tracking of vessel movements and monitoring of vessel activities through the use of vessel monitoring systems (VMS), onboard observers and aerial or at-sea surveillance activities. The physical inspection of catch, gear and documentation may also be carried out to ensure that fishing vessels are abiding by all applicable laws. Investment into enforcement activities and available resources varies greatly from country to country however, and for this reason there are some countries where IUU activity is much more prevalent than elsewhere, such as in parts of West Africa (Bray, 2001; Agnew et al., 2009).

Countries also take on responsibility for combatting IUU fishing activities through their role as flag States, in monitoring and enforcing relevant legislation upon all vessels that are registered under their flag. This responsibility has commonly been known as "flag State responsibility" and has been an issue of particular concern to international non-governmental organizations (NGO), governmental organizations, and to the EU (Bray, 2001; EC, 2008). Flag State responsibility is an issue that has been notoriously difficult to tackle, because of the complex nature of international maritime governance. Outside of coastal EEZs, flag States have the authority over the activities of vessels and the boarding of a foreign vessel without permission from the vessel's flag State is a violation of that country's sovereign rights under international law (UN, 1982). As mentioned previously, the EU has recently implemented a strategy whereby flag States are recognized for their non-cooperation with efforts to stem IUU activity. These countries are initially issued a warning from the EC, then provided with an opportunity to make improvements. If they fail, vessels that are registered under the flags of these States are not allowed to sell fish to any EU Member State (EC, 2008). At present, it is difficult to say what impact this strategy has had on decreasing IUU fishing activity globally, though flag States that in the past were recognized by the EC for their poor performances, have since improved their efforts, removing themselves from the EC's list of countries under watch. Belize is an excellent example of a country that has made significant improvements to their systems of fisheries governance after being warned and sanctioned by the EU. Speaking at the Ninth International Chatham House Forum on IUU Fishing held in London in February 2016, Belize's Ambassador to the EU, Dr. Dylan Vernon (2016) reported that in addition to improving their status as a more responsible fishing nation, as a result of increased licensing fees and fines, revenues to the government have increased by approximately 20%.

The identification and black listing of individual vessels that have been observed or apprehended for IUU fishing is a strategy that has been applied by national governments, the EU, RFMOs and NGOs (e.g., see www.iuu-vessels. org and www.blacklist.greenpeace.org). Consequences associated to the listing of vessels are varied depending on the listing authority, although always included is the public shaming of said vessels that results from simply being listed. Excluding vessels solely featured on NGO IUU fishing vessel lists, there is typically a requirement of States that fall under the jurisdiction of the listing authority to not allow fish caught by these vessels to be landed at ports or sold into markets that fall under their jurisdiction (e.g., EC, 2008). In 2013, the International Criminal Police Organization (INTERPOL) launched Project Scale, with the intention of supporting member countries in identifying, deterring and disrupting transnational fisheries crime. Through this project, INTERPOL has recently started issuing so-called "purple notices" for fishing vessels, providing and/or seeking information on modi operandi, objects, devices and concealment methods used by the criminals associated to these vessels. As of February 2016, purple notices have been issued for a total of nine fishing vessels (see www.interpol.int/INTERPOL-expertise/Notices/ Purple-notices-%E2%80%93-public-versions).

As IUU fishing is a problem affecting fisheries around the world, combatting IUU fishing is a global challenge that will require coordination of efforts internationally and cooperation amongst different countries and stakeholders (Doulman, 2000). An example of where in-depth collaboration between diverse stakeholders has been successful in addressing IUU fishing is in the Southern Ocean, where both Patagonian toothfish (*Dissostichus eleginoides*) and endangered albatross (Family Diomedeidae) (as by-catch) were being heavily impacted by IUU fishing activities during the 1990s. The substantial reduction in IUU fishing (estimated at >90%) has been attributed to international political pressure, and the development of tools that facilitated collaboration such as a catch documentation scheme and an IUU vessel black list (Osterblom et al., 2015). In addition, actions taken by non-state actors (licensed fishing companies and environmental NGOs) and policy measures adopted by individual states have been credited for leading to the subsequent development of impactful measures by CCAMLR, the RFMO mandated with the authority to govern fishing activities in the Southern Ocean (Osterblom et al., 2015). Lastly, it has been shown that toothfish caught from different stocks with different standards of management can be genetically differentiated (Marko et al., 2011). Thus, DNA testing can also potentially have a useful role in the regulation of fisheries in the Southern Ocean through detecting mislabeling in markets of fish falsely labeled as being caught from the more sustainably managed stocks (Marko et al., 2011).

A number of international agreements have emerged over the past few decades that specifically address IUU fishing. As mentioned previously in the IUU Fishing Defined section, in 2001, the UN adopted a voluntary instrument (the IPOA-IUU), that provided Member States with comprehensive measures for preventing, deterring and eliminating IUU fishing, and encouraged States to implement these directly, or indirectly (FAO, 2001). In 2009, the FAO adopted a legally binding instrument (the PSMA), which has just recently come into force. The main purpose of the PSMA is to prevent, deter and eliminate IUU fishing through the implementation of robust Port State Measures (FAO, 2009a). Under this legislation, countries take on responsibility for combatting IUU fishing activities through their role as port States, monitoring and controlling activities that are carried out in their ports through inspections, detentions and, when appropriate, denial of entry. Some port States have been criticized for failing to prevent IUU caught fish from entering the global seafood market and, much like "FoCs," "ports of convenience" are also known to exist, where fishing vessels looking to land catch without detailed documentation may find it easier to gain access to markets (Petrossian et al., 2014).

Alternative strategies for addressing IUU fishing can also potentially be found outside of the more traditional sphere of direct policy reform. Those who engage in IUU fishing activities are motivated through economic incentives and if the benefits of IUU fishing outweigh the costs, IUU activity will continue (Schmidt, 2005; Gallic and Cox, 2006; Sumaila et al., 2006). Both tangible and intangible motivations may influence individuals' decisions about whether or not to comply with fishing regulations, and these may include both moral obligations and social influence (Sutinen and Kuperan, 1999). Thus, the costs associated with the risk of apprehension for illegal fishing, which may for example include monetary fines and/or social consequences, should be considered along with the expected benefits when devising strategies for improving compliance (Sumaila et al., 2006).

IMPACTS OF ILLEGAL, UNREPORTED AND UNREGULATED FISHING ON THE SEAFOOD INDUSTRY

A study from 2009 estimated that the total value of current illegal and unreported fishing losses worldwide are between US $10 and US $23.5 billion annually, representing between 11 and 26 million tonnes of fish (Agnew et al., 2009). In the most recent time periods considered, illegal and unreported fishing was estimated to be present in all regions of the world, with the largest presence estimated in the Eastern Central Atlantic, off the coast of West Africa (Agnew et al., 2009). For the year 2011, it has been estimated that illegal and unreported catches represented between 20% and 32% by weight of the wild-caught seafood imported into the US (Pramod et al., 2014). Given these estimates, it is clear that IUU fishing is pervasive and a relevant issue for seafood markets worldwide.

As mentioned previously, IUU fishing can negatively affect the environment, law-abiding fishers and society. Direct consequences to the environment may include damage to habitats when harmful gear is used, and the depletion of targeted or non-targeted (retained as by-catch or discarded) fished populations. IUU fishing activities may thus influence the sustainability of a fishery or even the vulnerability of stocks to commercial extinction (Doulman, 2000). The global seafood industry is the commercial extension of both the international fishing and aquaculture industries, and so threats to the sustainability of wild fisheries also threaten the economic viability of commercial stakeholders within the seafood industry (Doulman, 2000). Law-abiding fishers suffer reduced fishing opportunities when their targeted fish stocks are also targeted by IUU fishing vessels (Doulman, 2000). Costs for these fishers may increase (e.g., through increased fuel costs from longer fishing trips) and profits may decrease. Fishers are also financially impacted when product mislabeling occurs as excess mislabeled fish within the market may suppress the interplay of supply and demand, preventing them from getting fair market value for their catch. The extent and reach of short-term economic losses from IUU fishing activity depends on where the IUU caught fish enters into the seafood supply chain as laundered catch. However, as stated previously, long-term economic losses could be borne by all stakeholders within the seafood industry that stand to be affected by depleted fish stocks, the reduced availability of seafood and altered seafood prices. For some developing countries, IUU fishing may threaten the livelihoods of small-scale, artisanal fishers and even the food security of coastal regions that rely heavily on fish for dietary protein (Doulman, 2000). This is an issue of particular concern in West African countries where IUU fishing has a strong presence (Agnew et al., 2009).

Indirectly, IUU fishing may affect the health of marine ecosystems which may have unpredictable consequences on other commercially important fish stocks. IUU fishing may also indirectly threaten the health of consumers (Jacquet and Pauly, 2008). When IUU caught fish is laundered into the global seafood market, if it is mislabeled to conceal its true identity or origin, the information that ultimately reaches the consumer regarding the product they consume would be incorrect. If the species recorded on the product label is incorrect, consumers may be at risk of being poisoned, or having an allergic reaction (Ling et al., 2009; Cohen et al., 2009). If the date of capture is incorrect, assumptions regarding how long seafood can safely be consumed before spoiling may be dangerously misinformed. Lastly, if information concerning catch location is incorrect, consumers may be unknowingly exposed to toxic and/or biohazardous contaminants if the seafood was caught in polluted waters (Jacquet and Pauly, 2008). Species substitution can also pose a similar risk, as some commercially caught species accumulate higher levels of toxins in their tissue than others depending in part on their trophic position within marine food webs (Tibbetts, 2003; Lowenstein et al., 2010).

Species substitution and the mislabeling of seafood also reduces the power that consumers have in their ability to make informed choices to drive down demand for, and alleviate fishing pressure on threatened species (Jacquet and Pauly, 2008). This would be an issue of particular concern in instances where products labeled as "sustainable" options are actually mislabeled products originating from overfished or depleted stocks. Evidence of this mislabeling strategy has been uncovered within Irish and UK markets, where products labeled as "responsibly sourced" Pacific cod (*Gadus macrocephalus*) were genetically identified as Atlantic cod (*Gadus morhua*) (Miller et al., 2012b). Stocks of Atlantic cod have historically been overfished on both sides of the Atlantic and because of this, fishery closures and strict quota limits have been enforced (Miller et al., 2012a). Thus it is possible, if not likely, that IUU caught fish, either caught over quota or from closed areas, are on occasion labeled as imported fish to disguise their origins and conceal IUU activity within Atlantic waters (Miller et al., 2012b).

Consumer-driven change can only ever be possible if consumers have access to accurate information and that accurate information can only ever be guaranteed if species substitution and seafood mislabeling are effectively dealt with within seafood supply chains. In the absence of action on this issue, IUU fishing will continue to be facilitated, resulting in sustained or increased short- and long-term losses to all stakeholders within the seafood industry.

REFERENCES

Agnew, D.J., Pearce, J., Pramod, G., et al., 2009. Estimating the worldwide extent of illegal fishing. PLoS One 4, e4570.

Birnie, P., 1993. Reflagging of fishing vessels on the high seas. Review of European Community & International Environmental Law 2, 270–276.

Bray, K., 2001. A Global Review of Illegal, Unreported and Unregulated (IUU) Fishing. Food and Agriculture Organisation (FAO), Rome, Italy.

CCAMLR, 1980. The Convention for the Conservation of Antarctic Marine Living Resources. The Commission for the Conservation of Antarctic Marine Living Resources, Hobart, Australia.

Cohen, N.J., Deeds, J.R., Wong, E.S., et al., 2009. Public health response to puffer fish (Tetrodotoxin) poisoning from mislabeled product. Journal of Food Protein 72, 810–817.

Doulman, D., 2000. Illegal, Unreported and Unregulated Fishing: Mandate for an International Plan of Action. Food and Agricultural Organisation (FAO), Rome, Italy.

EC, 2008. Council Regulation (EC) No 1005/2008 of 29 September 2008 establishing a Community system to prevent, deter and eliminate illegal, unreported and unregulated fishing, amending Regulations (EEC) No 2847/93, (EC) No 1936/2001 and (EC) No 601/2004 and repealing Regulations (EC) No 1093/94 and (EC) No 1447/1999. Official Journal of the European Union L 286, 1–32.

EJF, 2013. Transhipment at Sea: The Need for a Ban in West Africa. Environmental Justice Foundation, London.

FAO, 2001. International Plan of Action to Prevent, Deter and Eliminate Illegal, Unreported and Unregulated Fishing. Food and Agriculture Organisation (FAO), Rome, Italy.

FAO, 2009a. Agreement on Port State Measures to Prevent, Deter and Eliminate Illegal, Unreported and Unregulated Fishing. Food and Agriculture Organisation (FAO), Rome, Italy.

FAO, 2009b. Expert Consultation on Flag State Performance. FAO, Rome, Italy.

Flothmann, S., von Kistowski, K., Dolan, E., et al., 2010. Closing loopholes: getting illegal fishing under control. Science 328, 1235–1236.

Gallic, B.L., Cox, A., 2006. An economic analysis of illegal, unreported and unregulated (IUU) fishing: key drivers and possible solutions. Marine Policy 30, 689–695.

Garcia-Vazquez, E., Perez, J., Martinez, J.L., et al., 2011. High level of mislabeling in Spanish and Greek hake markets suggests the fraudulent introduction of African species. Journal of Agricultural and Food Chemistry 59, 475–480.

Griggs, L., Lugten, G., 2007. Veil over the nets (unravelling corporate liability for IUU fishing offences). Marine Policy 31, 159–168.

IMO, 1974. International Convention for the Safety of Life at Sea (SOLAS). International Maritime Organization.

Jacquet, J.L., Pauly, D., 2008. Trade secrets: renaming and mislabeling of seafood. Marine Policy 32, 309–318.

Ling, K.H., Nichols, P.D., But, P.P.-H., 2009. Fish-induced keriorrhea. Advances in Food and Nutrition Research 57, 1–52.

Lowenstein, J.H., Burger, J., Jeitner, C.W., et al., 2010. DNA barcodes reveal species-specific mercury levels in tuna sushi that pose a health risk to consumers. Biological Letters pii:rsbl20100156.

Marko, P.B., Nance, H.A., Guynn, K.D., 2011. Genetic detection of mislabeled fish from a certified sustainable fishery. Current Biology 21, R621–R622.

Miller, D.D., Clarke, M., Mariani, S., 2012a. Mismatch between fish landings and market trends: a western European case study. Fisheries Research 121–122, 104–114.

Miller, D., Jessel, A., Mariani, S., 2012b. Seafood mislabelling: comparisons of two western European case studies assist in defining influencing factors, mechanisms and motives. Fish and Fisheries 13, 345–358.

Miller, D.D., Mariani, S., 2010. Smoke, mirrors, and mislabeled cod: poor transparency in the European seafood industry. Frontiers in Ecology and the Environment 8, 517–521.

Miller, D.D., Sumaila, U.R., 2014. Flag use behavior and IUU activity within the international fishing fleet: refining definitions and identifying areas of concern. Marine Policy 44, 204–211.

Osterblom, H., Bodin, O., Sumaila, U.R., Press, A., 2015. Reducing illegal fishing in the Southern Ocean: a global effort. Solutions 4, 72–79.

Petrossian, G.A., Marteache, N., Viollaz, J., 2014. Where do "undocumented" fish land? An empirical assessment of port characteristics for IUU fishing. European Journal on Criminal Policy and Research 1–15.

Pramod, G., Nakamura, K., Pitcher, T.J., Delagran, L., 2014. Estimates of illegal and unreported fish in seafood imports to the USA. Marine Policy 48, 102–113.

Schmidt, C.-C., 2005. Economic drivers of illegal, unreported and unregulated (IUU) fishing. The International Journal of Marine and Coastal Law 20, 479–507.

Sumaila, U.R., Alder, J., Keith, H., 2006. Global scope and economics of illegal fishing. Marine Policy 30, 696–703.

Sumaila, U.R., Lam, V.W.Y., Miller, D.D., et al., 2015. Winners and losers in a world where the high seas is closed to fishing. Scientific Reports 5.

Sutinen, J.G., Kuperan, K., 1999. A socio-economic theory of regulatory compliance. International Journal of Social Economics 26, 174–193.

Theilen, J.T., 2013. What's in a name? The illegality of illegal, unreported and unregulated fishing. The International Journal of Marine and Coastal Law 28, 533–550.

Tibbetts, J., 2003. Mercury in Japan's whale meat. Environmental Health Perspectives 111, A752.

Triantafyllidis, A., Karaiskou, N., Perez, J., et al., 2010. Fish allergy risk derived from ambiguous vernacular fish names: forensic DNA-based detection in Greek markets. Food Research International 43, 2214–2216.

UN, 1982. United Nations Law of the Sea. United Nations, New York, USA.

UN, 1995. The United Nations Agreement for the Implementation of the Provisions of the United Nations Convention on the Law of the Sea of 10 December 1982 Relating to the Conservation and Management of Straddling Fish Stocks and Highly Migratory Fish Stocks. United Nations, New York, USA.

Vernon, D., 2016. An Update on the EU IUU Regulation.

White, C., Costello, C., 2014. Close the high seas to fishing? PLoS Biology 12, e1001826.

Windward, 2014. AIS Data on the High Seas: An Analysis of the Magnitude and Implications of Growing Data Manipulation at Sea. Windward, Tel Aviv, Israel.

Wong, E.H.-K., Hanner, R.H., 2008. DNA barcoding detects market substitution in North American seafood. Food Research International 41, 828–837.

Section II

DNA-Based Analysis for Seafood Authenticity and Traceability

Chapter 5

An Introduction to DNA-Based Tools for Seafood Identification

Amanda M. Naaum, Robert H. Hanner
University of Guelph, Guelph, ON, Canada

As discussed in the previous chapters, species- and/or population-level identification are often critical components of determining seafood product authenticity, implementing traceability and ensuring sustainability. Ideally, morphological characteristics can be used to identify seafood species. Species-specific characteristics such as size and coloration may be used to differentiate species, often based on guides with photographs or drawings to aid in identification using descriptions of what can be observed visually. However, identifying fish in this way often requires expert training. Also, most of the distinguishing morphological characteristics are removed in even lightly processed food products. With seafood production possible even while at sea, this presents a problem in identification. Additionally, the large number of commercial species or potential substitutes (see Chapter 1) make the task even more difficult. Finally, the question of geographic origin is impossible to answer without a more sophisticated approach. There are a wide range of possible methods available, but DNA-based methods are some of the most useful and widely implemented to address these critical questions.

Identifying different species using their DNA is based on identifying different genetic mutations acquired during evolution. Genetic differences, or polymorphisms, like other characteristics used to differentiate species, become fixed over time in different lineages. These differences can then be used to tell groups of organisms apart, if you know where on the DNA sequence to look. Different regions of the genome evolve at different rates and can be targeted to tell species, populations or even individuals apart. Other methods including morphology, behavior and biochemical analyses focus on identification based on phenotypic manifestations of genetic information. These manifestations can have a large range in how they are expressed physically, and can be heavily influenced by the environment as well as the species, and on the life stage and sex of the organism. Using DNA analysis allows direct identification of the genotype, and overcomes many of the potential issues with phenotypic approaches. The molecules are extremely stable, and can often survive harsh

Seafood Authenticity and Traceability. http://dx.doi.org/10.1016/B978-0-12-801592-6.00005-X

processing conditions without being altered beyond recognition and the same DNA is found in nearly all cells in an organism.

The advances in DNA analysis over the last three decades have resulted in the development of a variety of different methods, all becoming increasingly economical, accurate and accessible. New methods continue to be developed, while existing methods become cheaper and simpler to use with improved instrumentation. Technology that not too long ago was restricted to research institutions is now becoming available to the general public, as evidenced by the emergence of biohacking spaces where anyone can do DNA science (Michels, 2014). High school students are contributing data to seafood market surveys using DNA barcoding (e.g., Naaum and Hanner, 2015), and the instrumentation for DNA testing is available through Kick-starter campaigns (e.g., Bento Lab and OpenPCR). This easy access to DNA testing is a testament to the robustness and usefulness of many of these methods, and their continued application to seafood identification is almost assured.

This chapter discusses the methods available for DNA-based authenticity testing of seafood, both established and emerging. This chapter acts as a primer for DNA analysis, covering the basics behind several of the key methods most commonly used for seafood testing. This preliminary overview of the different techniques provides some background on how each works and discusses some of the pros and cons of each method.

DNA EXTRACTION

DNA extraction, or accessing the DNA from a sample, is the first step for almost every method of DNA analysis. This requires lysing cells to release the DNA from the nucleus, or other DNA-containing organelles. For fresh samples, this may be enough for use in certain applications. For example, polymerase chain reaction (PCR) success can sometimes be achieved after sonication of tissue in liquid to release DNA. However, more commonly lysis is followed by steps to separate DNA from other cellular debris, or other similar molecules such as proteins and RNA. This can be achieved through steps like centrifugation and/or filtration, for example. The number and types of steps vary between different methodologies. One of the oldest, and most successful, methods for DNA extraction is the use of phenol–chloroform extraction. Although still used today, this method has several drawbacks, including the long processing time and use of potentially harmful chemicals. To circumvent these difficulties, several different commercial options for DNA extraction exist. Commercial kits also offer consistency between extraction from different samples and across labs.

The KAPA Express Extract kit (KAPA Biosystems) is one of the most rapid on the market, involving a short incubation and centrifugation protocol that results in PCR-ready DNA in 15 min. Qiagen Blood and Tissue kits (Qiagen) involve more steps, usually including a lengthy incubation for lysis; however, the resulting DNA is usually of better quality, and more stable for longer term storage. For particularly problematic samples that may contain PCR inhibitors, food-specific kits like the Promega Wizard Magnetic DNA Purification System for Food can be employed. Whatman FTA cards (Whatman) offer a portable option, suitable for simple collection of forensic samples, and often used in law enforcement (Moran, 2010). Samples can be spotted onto cards containing silica and microbial growth inhibitors, which lyse cells and bind DNA. The sample can be stored long term at room temperature, which simplifies transport or shipping of samples. When DNA needs to be analyzed, a punch from the card is taken and after a short series of incubation/washing steps used directly in PCR. Finally, some commercial kits for species identification may come with DNA extraction components incorporated. Examples include the InstantID real-time PCR (qPCR) kits (InstantLabs) and the Fish DNA Barcoding Kit from BioRad, the latter of which is designed for educational settings.

The costs, extraction time, and the quality and yield of the DNA obtained vary widely between the different kits, and some research or optimization is sometimes necessary to make the most appropriate selection for a given application. Methods for nucleic acid quantification and purity analysis are available to help determine extraction success. Spectrophotometric methods measure absorbance at different wavelengths of light, and provide not only a quantification of the DNA in the sample, but also a measure of purity from protein or other contaminants based on the ratios of absorbance at 260 and 280 nm (e.g., Nanodrop 2000 from Thermo Fisher Scientific). Fluorescence-based DNA quantification is also possible (e.g., Qubit 3.0 Fluorometer from Thermo Fisher Scientific). Although these instruments do not provide a measure of purity, they can give more accurate quantification for samples with lower DNA concentration, which may be observed for some processed seafood. For samples from mixed products, sampling methods must ensure uniform sampling from the mixture, and this is another factor when choosing and/or modifying a commercial DNA extraction kit. For example, when sampling a can of tuna, a homogenization of a significant portion of the contents will give the most accurate analysis of the potential species contained in the product. The best extraction protocol is usually identified by researching published methods and trial and error. This must be assessed on a case-by-case basis.

GEL ELECTROPHORESIS

Gel electrophoresis is a common technique for visualizing DNA, and is a frequent part of DNA analysis for several identification methods. It can be used as a quality control measure to determine that a PCR product has been generated as simple presence/absence. This is often a step before DNA sequencing to confirm success of the preliminary PCR reaction, and that a single fragment of the expected size was amplified, confirming the specificity of the primers. It is also used to separate out DNA extract before PCR is conducted to determine what size fragments are present. A current is passed through a gel, and since DNA is negatively charged, it moves towards a positive electrode. Smaller fragments travel through the gel, which is usually made from agarose or acrylamide, more quickly, whereas larger fragments move more slowly due to their size. Using a DNA ladder, a solution of DNA of varying known size fragments, the size of a fragment from an unknown sample of PCR product or DNA extract can be estimated.

Samples are loaded onto a gel of appropriate composition and pore size, controlled by the percentage of agarose or acrylamide added to the buffer when the gel is cast. Higher percentage gels have smaller pores and increased resistance to DNA traveling through, whereas lower percentage gels have larger pores and less resistance. As a result, higher percentage gels are generally used to separate out smaller fragments because they provide greater resolution for smaller fragments. To visualize the DNA, a DNA-binding dye is incorporated into the gel. These dyes can vary, but fluorescence when viewed under UV light, revealing bands on the gel where DNA is present. Precast gels (e.g., E-gels, Thermo Fisher Scientific) can be purchased commercially, and are ideal for high-throughput settings. However, gels can also be relatively easily made, which is much more cost effective if they are used infrequently, or if budgetary constraints require it.

POLYMERASE CHAIN REACTION

PCR is probably the most revolutionary advancement in molecular biology. Kary Mullis won a Nobel Prize in 1993 for conceiving this method for generating copies of a target gene region in vitro. PCR allows a portion of DNA to be copied, making enough from even a small amount of starting sample to be easily visualized on a gel, or used for other downstream methods or applications. The DNA is copied using simple cycles of heating and cooling. To start, any DNA in the sample is split, or denatured, by heating to 95°C. The reaction is then cooled to the primer annealing temperature, usually between 50 and 60°C. At this temperature, primers, very short pieces of DNA complimentary to the target strand, bind to the target DNA if it is present in the sample. Target-specific primers are designed based on known sequences of DNA and are designed to amplify a single target, or small group. Universal primers can also be used. These primers are designed to bind to conserved regions of DNA that are present in many different potential targets and can be used to amplify a target region from many different organisms. For example, the primers used for DNA barcoding of fish are universal, and amplify the marker region from most fish species (Ivanova et al., 2007).

Once the primers bind, the reaction is heated again to 72°C. At this temperature, DNA polymerase is activated, and begins to build a complimentary strand of DNA to match the single strand to which the primers are bound (the template). The polymerase incorporates free individual deoxynucleotides (dNTPS), the building blocks of DNA, to match the template strand. In this way, the target DNA is doubled with each cycle. The Howard Hughes Medical Institute has developed several animations related to DNA science, including a video of the steps in PCR (http://www.hhmi.org/biointeractive/dna-collection).

Restriction Fragment Length Polymorphisms

RFLP has been the most commonly implemented method for species discrimination in food and it starts with PCR amplification of a target region that will allow species discrimination. This is followed by digestion of the amplicon with restriction enzymes. These enzymes cut DNA at a specific sequence of nucleotides, generating fragments of DNA with different sizes from the amplified target region. Analysis via gel electrophoresis separates these segments on the basis of size, creating a restriction profile (Fig. 5.1). These profiles differ depending on the species, due to differences in DNA sequence, and can be compared to a database of known profiles or to a reference standard to identify the species present. Two major benefits of this method are that the equipment required is easy to access and use, and that experimental design requires minimal upfront work as previous knowledge of underlying nucleotide sequences is not always necessary if universal primers are used for PCR. However, the process of sample preparation and analysis is lengthy and there are several post-PCR processing steps that increase the likelihood of error due to the failure of a specific step, or sample mix up, or of contamination. Also, issues with reproducibility call the accuracy of the method into question (Lockley and Bardsley, 2000). Inclusion of positive controls with each gel can improve results. In addition, electronic chips can be employed for more automated interpretation of results (see Chapter 7).

Real-Time PCR

Although almost two decades old, real-time PCR has only been introduced relatively recently as a tool for food authentication. This method takes species-specific PCR a step further by combining PCR amplification and detection in a single step, and by allowing quantification of a target. This property gives the method its abbreviation of "qPCR," for "quantitative PCR," though it can be used as a powerful identification tool without taking advantage of the potential for quantification. Each PCR reaction goes through a linear phase as the replication process begins, followed by a period of exponential growth. A plateau phase is then reached where amplification efficiency is reduced. In traditional PCR, post-PCR steps must be employed and therefore, analysis only takes place after the reaction is completed and has reached the plateau phase. In qPCR, the exponential phase of the reaction can be identified since the reaction is monitored for its complete duration. Fluorescence emission during the exponential phase is directly proportional to the DNA copy number therefore qPCR allows quantification in addition to identification. Using species-specific primers and/or probes makes it possible to use this method to authenticate the content of food products if sequence information is available

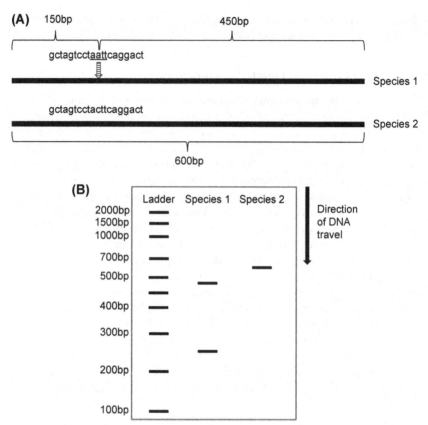

FIGURE 5.1   (A) In RFLP, enzymes cut DNA at a specific sequence; in this case, the enzyme EcoRI cuts at all AATT locations. In this way, sequences that contain AATT at a specific location can be differentiated from a sequence that does not using gel electrophoresis. Multiple enzymes or sites may be used together to tell species apart. (B) The result of gel electrophoresis from PCR-RFLP. The ladder of known fragment sizes is used to estimate the sizes of bands from test specimens. In this example, the profile for species 1 (containing the AATT cut site) shows two bands at 150 and 450 bp, since the longer segment of DNA has been cut. For species 2, a band of 600 bp, the full length of the longer segment, is observed. *PCR*, polymerase chain reaction; *RFLP*, restriction fragment length polymorphisms.

from which to design specific primers and/or probes. Various fluorescent reporter chemistries are available to carry out qPCR.

   The first variety is non-specific chemistries. These employ dyes that bind to all double-stranded DNA (dsDNA). SYBR Green, the most commonly used, emits more fluorescence when bound to dsDNA than when free in solution, allowing it to detect the presence of dsDNA. Non-specific chemistries are direct and simple, and the most cost-effective approach to qPCR. However, they are limited as they cannot differentiate signal from target dsDNA from non-specific products or primer dimer. To address this, a melt curve must be run for each

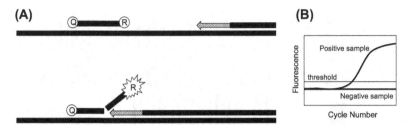

FIGURE 5.2 (A) Fluorescence is generated from a TaqMan probe when the reporter dye (R) is separated from the quencher (Q) during normal extension of template DNA during real-time PCR (qPCR). (B) The resulting fluorescence output is simple to interpret for presence or absence of the target. Samples with the target DNA present exhibit fluorescence that crosses a given threshold. A negative sample with no target DNA present will not exhibit a signal.

sample. Since different sized products have different melting temperatures, a melt curve allows the presence of multiple products of different sizes to be identified, if present, by recording the melting temperature for all products of the qPCR reaction. Non-specific chemistries are also limited in specificity as this is determined by only the primers, whereas probe-based chemistries also include a target-specific probe.

Probe-based chemistries are the alternative to the non-specific approach. TaqMan probes are the most commonly used probe-based chemistry. This method is based on the use of a probe with a quencher dye at one end and a reporter dye at the other end. When in close proximity, joined together on either end of the probe, the fluorescent signal from the reporter dye is masked by the quencher dye. During PCR amplification, the probe is cleaved as the complimentary DNA strand is generated, releasing the reporter dye and emitting fluorescence, which is measured by the instrument (Fig. 5.2). Different reporter dyes, emitting fluorescence at different wavelengths, can be used. Some instruments have multiple channels of detection that each monitors a specific range of wavelengths. By combining probes modified with dyes of different emission spectra, known as multiplexing, simultaneous detection and/or quantification of multiple targets in a single reaction can be achieved. Although more expensive than non-specific chemistries, probes add an additional level of specificity to a qPCR assay. They also reduce the assay time, as a melt curve need not be generated or analyzed.

Some qPCR platforms, although generally more costly than traditional PCR thermal cyclers, can also be portable, allowing on-site analysis. In addition, interpretation of results is simple and analysis is very rapid in comparison to other methods. Finally, since there is no post-PCR processing, the possibility of contamination from previous PCR products and the time required for analysis are greatly reduced.

DNA Sequencing

The dideoxy, or chain termination sequencing method is often referred to as Sanger sequencing, named after Frederick Sanger who was awarded a Nobel Prize in 1980

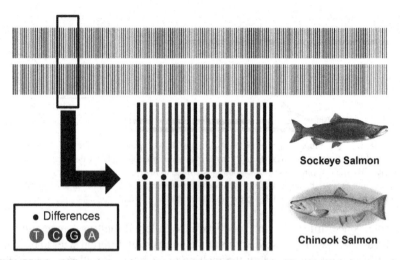

FIGURE 5.3   Differentiation of species using DNA sequencing is achieved by directly comparing sequences. Identifying these sequence differences, forms the basis for other DNA-based identification methods, for example, allowing species-specific primer design, or identifying SNPs. DNA sequence knowledge is a critical first step in almost all DNA identification methods. Identification of an unknown sample can be made using DNA sequencing directly, by comparison to a database of sequences with known identity. *SNPs*, single-nucleotide polymorphisms.

for the development of the technique. DNA sequencing provides the most information about a sample, as it reveals the nucleotide sequence of an entire portion of DNA (Fig. 5.3). It has also been used for species authentication in food. In this approach, a small region of DNA up to 1000 bp is amplified using PCR, and then a sequence of the amplicon is obtained using Sanger sequencing. The sequence is then compared to a database of sequences to determine the species. Rather than simply authenticating a sample from a small group of anticipated possibilities, DNA sequencing can be used to identify the species against all known sequences in a database. DNA sequencing has been applied to identify species in food products directly; however, this method is not generally suitable for all food products as the ability to detect species in mixtures is limited. Additionally, the region used may be too long to recover from highly processed samples unless a smaller fragment is targeted (e.g., Shokralla et al., 2015). Despite these limitations, DNA sequencing is a valuable tool for identification and authentication of unknown samples in many applications. It is also of key importance in the design and success of the aforementioned methods of species identification, particularly in the generation of reference databases used for sequence comparison or for PCR assay design. Chapter 6 covers DNA sequencing for seafood species identification in detail.

Population Identification

The two most common approaches for the identification of population of origin are based on analysis of microsatellites or single-nucleotide polymorphisms (SNPs). Microsatellites are segments of a repeated set of nucleotides (2, 3 or 4) that vary in the number of repeats. Microsatellites are resolved by electrophoretic methods that allow visualization of the number of repeats. By combining information on the number of repeats at various sites, populations can often be differentiated. SNPs are specific locations where a single nucleotide varies between individuals or groups. Combinations of these can also be used to tell populations apart using probabilistic inference. They are identified using direct sequence comparisons, but then routinely tested for using electrophoretic methods. The specific means of analysis and identification from data generated using microsatellite or SNP markers, including the required statistical analysis of data, can be variable. Both methods are much more labor intensive than those for species identification, and require a great deal of upfront knowledge and expertise to design and use. However, they give the most comprehensive information about a sample, and can be critical in addressing specific questions in seafood authenticity, and particularly traceability, that cannot be more simply answered. Chapter 8 provides more detail on the use of these methods for population identification.

Choosing a Method

The choice of which analytical tool to use for DNA analysis depends on several factors. Primarily, the key reason for testing must be identified. Population-level differentiation would require different methods and markers than species-level identification. Methods like SNP analysis and microsatellites require a much more comprehensive biological understanding of the populations and much heavier sequence coverage of many more individuals from target populations than for species discrimination using DNA sequencing for example. Furthermore, a quantitative result may be required, or the testing method may need to be portable. Additional considerations should include the equipment and facilities available, cost, time required for analysis, expertise required, existence of commercial kits and the availability of appropriate reference material and sequences. Appropriate validation should be carried out before using any new method. This validation should include analysis of specificity and sensitivity of the method for the target species or population, closely related species or populations or more distantly related groups that may be confused with the target in some way. Validation should also be done on products after applicable processing steps, ideally spanning the range of possible processing for a sample, to ensure that the method selected performs as expected, or to set guidelines for differential analysis of samples that have been processed differently. Published guidelines can be helpful when designing and validating new protocols (e.g., Bustin et al., 2009; Naaum et al., 2015).

CHOOSING A MOLECULAR MARKER

Selecting which marker to use is as important as the analytical tools used to make an identification. In general, a marker is a characteristic that can be measured that provides some information about classification. It may be morphological or biochemical, for example. Molecular markers specifically describe DNA characteristic that can be used to identify some genetic trait. Any segment of DNA, combined with an analytical technique to characterize it, meets this definition. Currently, a handful of different markers have been targeted for DNA-based species identification in animals. Largely these are mitochondrial markers, selected for their presence in high copy number, lack of recombination and more rapid rate of molecular evolution, compared to nuclear DNA. This can be crucial for the analysis of processed or otherwise degraded samples where only trace amounts of DNA may be present. The DNA region selected to discriminate species directly affects the success of the methods discussed earlier. It must be variable enough to discriminate between species, yet exhibit low variability within the same species. In addition, the length of amplicon must be appropriate for the processing level of the product tested as longer fragments can be difficult to recover from highly processed seafood. Standardization of the region for species identification, coupled with broad taxonomic surveys, would aid in the development of diagnostic testing for food authentication by allowing tests to be more broadly applicable across a wide range of taxa. In cases where molecular species identification is needed, a compelling argument can be made for using DNA barcoding, a species identification method based on sequencing of a short, standardized gene region, as the standard for food analysis because it has the most densely populated reference library.

The DNA barcode region for animals is 648 bp of the mitochondrial *cytochrome c oxidase* subunit one (COI) gene (Hebert et al., 2003). This region, like others selected for species identification, generally exhibit low levels of variability within a species, and higher levels of variability between species. DNA barcode sequences available on the Barcode of Life Data System (BOLD; www.barcodinglife.org, Ratnasingham and Hebert, 2007) are accompanied by data that can be crucial to successful design of assays for species identification. The workbench available through BOLD supports addition of metadata like images and GPS coordinates, and supports third-party annotation. Errors in common nucleotide databases, like GenBank (Harris, 2003), can result in incorrect identification when using sequencing, and also lead to false positives and false negatives if used in the development of other identification assays. Because it is curated, the BOLD database is a useful source of high-quality sequences and corresponding metadata for the development of DNA-based assays for species differentiation in food products. Many of these sequences are generated from specimens identified by expert taxonomists, reducing the likelihood or erroneous identifications. A morphological voucher specimen held in a museum collection also makes it possible to re-confirm identifications of reference specimens at a later date and enhances the evidentiary value of the sequence. BOLD can provide comparative sequence data from the same genetic marker for a range of targets for assay development, and also the option to include individuals from varied geographic range when selecting sequences from which to develop an assay.

For sequences to be considered a true "BARCODE" sequence, BOLD requires raw trace files to be submitted. When comparing sequences for species identification using genetic distance, one or two errors in a DNA sequence may not make a difference in accurate identification. However, when developing species-specific primers or probes, a single-nucleotide difference can affect results, causing false positives or false negatives. The ability to view trace files avoids inclusion of artificial haplotypes generated from errors in sequence assembly. Standardization of the gene region used allows a wider range of taxa, and a larger number of target individuals from a wider geographic range to be included in assay design without upfront sequencing costs associated with obtaining specimens and generating new sequences each time an assay is developed.

DNA barcoding has been applied across a wide range of taxa and results are publicly accessible in BOLD. Many of the most commercially important species have already been targeted, and the continued growth of the database of DNA barcode sequences will increase the method's robustness to the identification of adulterated foods. In fact, the use of DNA barcoding for food authenticity has been reviewed (Galimberti et al., 2013), and the method has been adopted as the official regulatory test for seafood in the United States (Handy et al., 2011).

Although useful for species identification in many cases, the DNA barcoding region is not always suitable particularly for closely related species that may hybridize. Markers should be selected based on the level of resolution that is necessary and on what groups need to be differentiated. A region with the appropriate rate of evolution is critical and a marker that easily differentiates species may not be appropriate to discriminate higher taxonomic categories, such as families. The evolutionary history of the target should be considered, as should availability of information about not just the target species or other group, but also those from which the target must be differentiated.

Limitations of DNA Analysis

Although outside the scope of this review, other DNA-based methods for species authentication that have been used to a lesser extent include biosensors and microarrays in addition to next-generation sequencing techniques, which may begin to be applied more commonly to issues relating to food authenticity. DNA can be used for more than authenticating species as some instances of food adulteration relate to the geographic origin of the product. In these instances, methods such as Amplified Fragment Length Polymorphism (AFLP), SNP analysis and microsatellites, mainly used to differentiate populations, may be used. Several DNA-based methods, including those discussed earlier, are commonly used for the identification of breeds/varieties/cultivars and genetically modified organisms (GMOs). These are not questions of authenticity at the species level, but still benefit from DNA analysis for authentication. These methods often require a large amount of upfront work to locate and select regions of DNA that will allow differentiation between populations or varieties.

There are other instances of adulteration that cannot be detected by DNA analysis. Authentication of a product as wild caught or organic, for example, cannot normally be carried out using DNA. Other analytical methods, such as analysis of stable isotopes, have been suggested for a small number of these applications; however, they are often unable to clearly identify these attributes (Sforza, 2013). Currently, the best approach for establishing authenticity of these types of products is rigorous supply chain traceability.

Despite these limitations, DNA testing represents one of the most well-studied and most promising methods for the identification of seafood species and populations. With the right test and marker selection, DNA analysis of seafood can help industry, regulators and even consumers verify authenticity and combat IUU fishing, leading to better fisheries management. The later chapters will review the use of various DNA-based methods for species- and population-level identification as applied to seafood, and discuss advances and future trends in these applications, including emerging technologies like next-generation sequencing and isothermal amplification.

REFERENCES

Bustin, S.A., Benes, V., Garson, J.A., Hellemans, J., Huggett, J., Kubista, M., Mueller, R., Nolan, T., Pfaffl, M.W., Shipley, G.L., Vandesompele, J., Wittwer, C.T., 2009. The MIQE guidelines: minimum information for publication of quantitative real-time PCR experiments. Clinical Chemistry. http://dx.doi.org/10.1373/clinchem.2008.112797.

Galimberti, A., De Mattia, F., Losa, A., Bruni, I., Federici, S., Casiraghi, M., Martellos, S., Labra, M., 2013. DNA barcoding as a new tool for food traceability. Food Research International 50, 55–63.

Handy, S.M., Deeds, J.R., Ivanova, N.V., Hebert, P.D.N., Hanner, R., Ormos, A., Weigt, L.A., Moore, M., Yancy, H.F., 2011. A single laboratory validated method for the generation of DNA barcodes for the identification of fish for regulatory compliance. Journal of AOAC International 94, 201–210.

Harris, J.D., 2003. Can you bank on GenBank? Trends in Ecology & Evolution 18, 317–319.

Hebert, P.D.N., Cywinska, A., Call, S.L., DeWaard, J.R., 2003. Biological identifications through DNA barcodes. Proceedings of the Royal Society of London B: Biological Sciences 270, 313–321.

Ivanova, N.V., Zemlak, T.S., Hanner, R.H., Hebert, P.D.N., 2007. Universal primer cocktails for fish DNA barcoding. Molecular Ecology Notes 7, 544–548. http://dx.doi.org/10.1111/j.1471-8286.2007.01748.x.

Lockley, A.K., Bardsley, R.G., 2000. DNA-based methods for food authentication. Trends in Food Science & Technology 11, 67–77.

Michels, S., September 2014. What Is Biohacking and Why Should We Care? PBS. Accessed at: http://www.pbs.org/newshour/updates/biohacking-care/.

Moran, B., April 2010. DNA Sampling Made Easy. Forensic Magazine. Accessed at: http://www.forensicmag.com/articles/2010/04/dna-sampling-made-easy.

Naaum, A., StJaques, J., Warner, K., et al., 2015. Standards for conducting a DNA barcoding market survey: minimum information and best practices. DNA Barcodes 3 (1), 80–84. http://dx.doi.org/10.1515/dna-2015-0010.

Naaum, A., Hanner, R., 2015. Community engagement in seafood identification using DNA barcoding reveals market substitution in Canadian seafood. DNA Barcodes 3 (1), 74–79. http://dx.doi.org/10.1515/dna-2015-0009.

Ratnasingham, S., Hebert, P.D.N., 2007. BOLD: the barcode of life data System. Molecular Ecology Notes. 7, 355–364. www.barcodinglife.org.

Sforza, S., 2013. Food Authentication Using Bioorganic Molecules. DEStech Publications, Lancaster, Pennsylvania.

Shokralla, S., Hellberg, R.S., Handy, S.M., King, I., Hajibabaei, M., 2015. A DNA mini-barcoding System for authentication of processed fish products. Scientific Reports 5, 15894. http://dx.doi.org/10.1038/srep15894.

Chapter 6

Seafood Species Identification Using DNA Sequencing

Rosalee S. Hellberg[1], Sophia J. Pollack[1], Robert H. Hanner[2]
[1]Chapman University, Orange, CA, United States; [2]University of Guelph, Guelph, ON, Canada

INTRODUCTION

DNA sequencing is a widely used method for the identification of seafood species and is advantageous in its ability to obtain high-level genetic information (Rasmussen Hellberg and Morrissey, 2011; Rasmussen and Morrissey, 2008). Unlike some DNA-based methods that are designed to target a narrow group of seafood species, DNA sequencing can be used to identify a wide range of organisms through the use of "universal" polymerase chain reaction (PCR) primers that can amplify a standard gene region across diverse taxa. This is especially beneficial when considering the wide variety and high number of seafood species that are commercially available as well as the increasing levels of international trade of seafood in processed forms (Rasmussen Hellberg and Morrissey, 2011). Although DNA sequencing remains relatively costly and time consuming compared to some of the other DNA-based methods used in species identification, technological advances in the field, such as automation and rapid PCR cycling, have helped to alleviate many of these concerns (Rasmussen Hellberg and Morrissey, 2011). Furthermore, the genetic information revealed through DNA sequencing can be used to build extensive reference libraries as well as provide a basis for the design of species-specific assays for rapid identification of selected organisms.

Traditional DNA sequencing methods are based on Sanger sequencing, in which chain-terminating dideoxynucleotides are incorporated into sequences using DNA polymerase (Sanger et al., 1977). Advances in the field of molecular biology have greatly improved the efficiency of this method since its invention; however, there continue to be numerous steps involved. First, the target genetic region is amplified using PCR, then the resulting amplicons typically undergo a cleanup step, followed by attachment of dye-labeled dideoxynucleotides in the cycle sequencing reaction, a sequencing cleanup step, and then capillary electrophoresis to determine the identity and order of nucleotides in the sequence.

Seafood Authenticity and Traceability. http://dx.doi.org/10.1016/B978-0-12-801592-6.00006-1

Next, the raw data are assembled and edited to generate a consensus sequence. Typical sequencing-based methods for seafood species identification rely on a reference database for species identification, in which the nucleotide sequence of an unknown sample is compared to a collection of reference sequences representing a range of seafood species (Rasmussen Hellberg and Morrissey, 2011). Phylogenetic analysis can then be performed and genetic distances are calculated to determine whether the query sequence has a reasonable genetic match to any of the reference sequences.

Identification methods based on the combination of DNA sequencing with phylogenetic analysis and genetic distance calculations are typically referred to as forensically informative nucleotide sequencing (FINS) or DNA barcoding (Rasmussen Hellberg and Morrissey, 2011). These methods rely on sequence databases to find a species match between an unknown sample and a reference sample. While FINS preceded DNA barcoding, the latter is advantageous in that it provides a standardized, high-throughput workflow for the analysis of unknown samples and the collection of reference sample information (Borisenko et al., 2009; Kress and Erickson, 2012). This chapter reviews important points to consider when using DNA sequencing to identify seafood species, including the choice of genetic target, the category of seafood products and species targeted, the availability of reference libraries, and methods for identifying species based on sequence information. Additionally, the potential use of next-generation sequencing (NGS) for seafood species identification and current research into the application of DNA sequencing to detect seafood mislabeling on the commercial market is also discussed.

GENETIC TARGETS

To reliably identify seafood at the species level with DNA sequencing, an appropriate genetic target must be selected. The ideal genetic target for this application must be easily amplified across diverse taxa and possess a sequence that is relatively conserved within species but exhibits significant sequence divergence between species (Rasmussen Hellberg and Morrissey, 2011). Mitochondrial DNA (mtDNA) has several advantages over nuclear DNA (nDNA) for species identification purposes, including a higher copy number, a lack of sequence ambiguities from heterozygous genotypes, and a faster rate of mutation (Rasmussen and Morrissey, 2008). However, as will be discussed later, mtDNA is not reliable for the differentiation of species that have a history of hybridization. A number of mitochondrial genetic targets have been investigated for the purpose of seafood species identification, including cytochrome c oxidase subunit I (COI), cytochrome b (cyt b), and 16S rRNA. In general, the top two genetic targets for seafood species identification have been the protein-coding genes COI and cyt b. These genes have been shown to be ideal for species differentiation because they are generally conserved within species and divergent between species (Rasmussen Hellberg and Morrissey, 2011). Although the

16S rRNA gene has also proven useful for seafood species identification, it has a lower rate of divergence than the other two genetic targets, which typically show greater than 2% sequence divergence between species (Armani et al., 2015b; Hebert et al., 2003b). Furthermore, the occurrence of insertions and deletions in genes coding for ribosomal DNA, such as 16S rRNA, can complicate sequence alignments (Hebert et al., 2003a). In a comparison of genetic markers, Kochzius et al. (2010) reported that COI and cyt *b* were suitable for the identification of 50 European marine fish species, whereas 16S rRNA was unable to differentiate some flatfish and gurnard species. A study focused on the differentiation of oyster species (Family Ostreidae) reported overlapping intra- and interspecific genetic divergences for 16S rRNA, but not for COI, indicating the usefulness of COI as a molecular marker for oyster identification (Liu et al., 2011).

Numerous studies have been published reporting the utility of cyt *b* in the differentiation of seafood species (reviewed in Rasmussen and Morrissey, 2009). A commonly targeted fragment within this gene is a 464-bp region, for which universal primers have been developed (Kocher et al., 1989). This region has been utilized for the sequencing-based differentiation of seafood species such as gadoids (Calo-Mata et al., 2003) and scombroids (Unseld et al., 1995). However, more seafood species identification studies have targeted cyt *b* fragments that are shorter in length (Cutarelli et al., 2014; Huang et al., 2014; Lago et al., 2011). For example, Cutarelli et al. (2014) utilized a ~360-bp region of cyt *b* to differentiate commercial fish species on the Italian market. Interestingly, this genetic region allowed for identification of 16 out of 18 species analyzed, while COI was able to identify all species tested (Cutarelli et al., 2014).

Owing to the strong phylogenetic signal of COI, as well as the availability of robust universal primers, a ~650-bp region of this gene has been selected as the standard for DNA barcoding (CBOL, 2015; Hebert et al., 2003a). DNA barcoding is a system for species identification focused on the use of a short, standardized genetic region acting as a "barcode" in a similar way that Universal Product Codes (UPCs) are used by supermarket scanners to distinguish commercial products (Fig. 6.1). The methodology was advanced as part of the Fish Barcode of Life (Fish-BOL) Campaign (Ward et al., 2009), which is a global initiative to collect DNA barcodes for all species of fishes. Extensive research has been carried out examining the use of COI as a molecular marker for seafood species identification (Hubert et al., 2008; Kim et al., 2012; Landi et al., 2014; Steinke et al., 2009; Ward et al., 2005; Yancy et al., 2008; Zhang and Hanner, 2012) and it has been found to be capable of differentiating 93% of freshwater fish species tested and 97% of marine species tested (Ward et al., 2009). Detailed, step-by-step protocols have been published for building barcode libraries (Hubert et al., 2008; Steinke and Hanner, 2011) and DNA barcoding of both fishes (Weigt et al., 2012) and invertebrates (Evans and Paulay, 2012).

DNA barcoding based on COI has been adopted by the US Food and Drug Administration (FDA) for use in the identification of regulatory fish samples (Handy et al., 2011a; Handy et al., 2011b; Yancy et al., 2008) and efforts

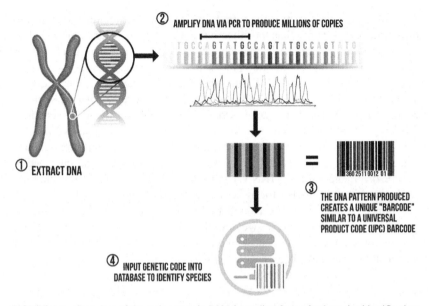

② AMPLIFY DNA VIA PCR TO PRODUCE MILLIONS OF COPIES

① EXTRACT DNA

③ THE DNA PATTERN PRODUCED CREATES A UNIQUE "BARCODE" SIMILAR TO A UNIVERSAL PRODUCT CODE (UPC) BARCODE

④ INPUT GENETIC CODE INTO DATABASE TO IDENTIFY SPECIES

FIGURE 6.1   Overview of the major steps in DNA barcoding for seafood species identification. First, the DNA is extracted from a seafood sample, then the DNA barcode region undergoes PCR amplification and DNA sequencing. The resulting DNA barcode sequence can be searched against a database, such as the Barcode of Life Database (BOLD), to find reference sequences that match the query sequence. *Kyle Kane; database image obtained from* http://www.boldsystems.org/.

are underway to expand current protocols to include decapod crustaceans, primarily shrimp, crab, and lobster, sold on the commercial market (Deeds et al., 2014). The inclusion of commercial mollusks, such as shellfish, squid, and octopus, in these protocols has been identified as a likely area of focus for the future. In an example of the utility of COI DNA barcoding in regulatory applications, Cohen et al. (2009) described a case in the United States in which puffer fish was fraudulently labeled and sold as monkfish, resulting in serious illness in consumers. DNA barcoding was used during the investigation to identify the product involved in the illness as belonging to the puffer fish family (Tetraodontidae). The Brazilian government has also begun using COI DNA barcoding to identify instances of seafood species mislabeling on the commercial market (Carvalho et al., 2015). In a study describing these efforts, 7 out of 30 products collected from supermarkets, fishmongers, and restaurants were found to be mislabeled and multiple products were found to contain vulnerable or endangered species (Carvalho et al., 2015). Establishments found to be selling fraudulent products were officially notified by the government and received financial penalties. Owing to the standardization of COI DNA barcoding protocols, the use of this method as a means to detect instances of seafood mislabeling may be adopted by more government institutions worldwide in the coming years.

Although cyt *b* and COI have been highly successful for use in species identification, they continue to have several limitations. For instance, these protein-
coding genes sometimes do not exhibit sufficient variation to differentiate
closely related species based on genetic distances and phylogenetic analyses.
This has been observed in studies focusing on identification of tuna species,
in which case alternative markers, such as the mitochondrial control region or
D-loop, have been found to be more appropriate (Cawthorn et al., 2011; Viñas
and Tudela, 2009). An alternative approach to sequence analysis, referred to as
character-based analysis, has also been examined as a supplementary method
for use with closely related species and is discussed later in the following section. Another issue of concern is that mtDNA is inherited only along the maternal line and, therefore, cannot be used to reliably identify individuals that are
the result of hybridization between species (Rehbein, 2013). In these cases, a
nuclear gene target must be used, such as the internal transcribed spacer (ITS1),
5S rRNAs, or rhodopsin (Sevilla et al., 2007). For example, introgression has
been observed among some species of tuna (e.g., *Thunnus alalunga* and *Thunnus thynnus*), resulting in the potential for misidentification based on mtDNA
alone (Viñas and Tudela, 2009). Therefore, a region of ITS1 has been identified
as a supplemental genetic marker to be used in differentiating these species
(Chow et al., 2006; Viñas and Tudela, 2009) while rhodopsin has been used as
a supplemental nDNA target for other species (e.g., in the aquarium trade, see
Collins et al., 2012).

Another important consideration when selecting an appropriate genetic target
for species identification is the level of processing that the seafood products have
undergone. In heavily processed seafood products, such as canned samples, the
DNA is typically degraded into fragments less than 350 bp (Rasmussen Hellberg
and Morrissey, 2011) making it an unsuitable template for the PCR amplification
of larger fragments commonly targeted using universal cyt *b* or COI primers. To
allow for sequencing-based species identification in these cases, shorter regions
of standardized gene targets (i.e., 100–300 bp), sometimes referred to as "mini-
barcodes," can be utilized (Hajibabaei and McKenna, 2012; Hajibabaei et al.,
2006). Potential mini-barcode regions within COI have been identified for the
simultaneous differentiation of a wide range of animal species (Hajibabaei et al.,
2006; Meusnier et al., 2008), as well as for the differentiation of specific groupings
of seafood species (Armani et al., 2015a; Armani et al., 2013; Bhattacharjee and
Ghosh, 2014; Santaclara et al., 2014). Hajibabaei et al. (2006) identified several
mini-barcode regions (109–218 bp) within the full-length COI DNA barcode that
could potentially be used for species differentiation. An in silico analysis of these
regions using a dataset of 204 Australian fish species (Ward et al., 2005) revealed
similar intra- and interspecies divergence levels as compared to the full-length
COI DNA barcode, indicating that species identification with a carefully selected
mini-barcode region could be used to address highly processed samples with
degraded DNA. Along these lines, a COI mini-barcoding system has been developed for fish species identification in processed products (Shokralla et al., 2015).

Six mini-barcode primer pairs were designed based on in silico analysis of DNA barcodes from fish species in the FDA Seafood List and the primers were then tested against 44 processed fish products. Overall, the mini-barcoding system showed increased sequencing success as compared to the full-length DNA barcoding, with identification to the genus level for 93% of the products and identification to the species level for 86% of the products.

Short stretches of DNA with applications in sequencing-based differentiation of heavily processed seafood have been identified for other genetic regions, including cyt *b* (Dalmasso et al., 2013; Espiñeira et al., 2009; Huang et al., 2014; Lago et al., 2011; Unseld et al., 1995), 12S rRNA (Ardura et al., 2010), and 16S rRNA (Armani et al., 2015b). However, many of these genetic regions have only been examined for the purpose of differentiating certain species groups, such as tuna species, rather than simultaneous differentiation of a wide range of seafood species. An exception to this is a study by Armani et al. (2015b), in which universal primers were designed to target 118- and 213-bp fragments of the 16S rRNA gene across multiple families of fish (Clupeidae, Engraulidae, Salangidae, and Scombridae). While the primers were successfully used for species identification in most of the samples analyzed, there were a few products that could only be identified to the genus level. Overall, further studies are needed to determine the optimal mini-barcode regions and primer sets to be used in identifying seafood species through DNA sequencing.

REFERENCE LIBRARIES

An essential part of species identification using DNA sequencing is the use of a reliable and comprehensive reference library. Clearly, in order for an unknown sample to be identified at the species level, the library must contain reference sequences representing that species. Development of reference libraries for seafood species identification is particularly challenging due to the high number of species that exist globally (Rasmussen Hellberg and Morrissey, 2011). Furthermore, the similar appearance of many seafood species gives rise to the possibility of misidentification and erroneous sequence information. A large amount of sequence data is available in the public database GenBank (http://www.ncbi. nlm.nih.gov/genbank/), which can be queried with the Basic Local Alignment Search Tool (BLAST). However, GenBank is not standardized specifically for seafood species identification and it has been criticized as being susceptible to issues such as inconsistent terminology and missing information (Harris, 2003; Rasmussen and Morrissey, 2008). Furthermore, there is no system in place to link sequence information with authenticated voucher specimens. On the other hand, the Fish Barcode of Life Initiative (FISH-BOL; www.fishbol.org) is a reference database established specifically for fish species and it is focused on the COI DNA barcode (Ward, 2012; Ward et al., 2009). The mission of FISH-BOL is to obtain standardized reference sequences for all fishes based on taxonomically verified voucher specimens. Sequences gathered through FISH-BOL are publicly archived in a number of databases, including GenBank, the DNA Data Bank of Japan (DDBJ), and the European Molecular Biology Laboratory (EMBL) database. Currently, FISH-BOL contains over 100,000 DNA barcode sequence records representing some 11,000 species. Similarly, the Marine Barcode of Life Initiative (MarBOL; http://www.marinebarcoding.org/) was developed to facilitate the ability to identify marine life through COI DNA barcodes. MarBOL is a joint initiative of the Consortium for the Barcode of Life (CBOL) and the Census of Marine Life (CoML) and has barcode records for close to 40,000 specimens representing over 6000 species, including both vertebrates and invertebrates. A user-friendly search engine specifically dedicated to identification of query sequences based on COI DNA barcodes is available online at the Barcode of Life Database (BOLD; http://www.boldsystems.org/index.php/ IDS_OpenIdEngine). This resource has the option of searching against all available barcode records (minimum length 500 bp) as well as narrowing the search to all available barcode records with a species-level designation or focusing on barcodes from published projects only.

A shortcoming of reference libraries such as those described earlier is the possibility of erroneous sequence information as a result of taxonomically misidentified specimens, specimen mix-ups, or sample contamination (Ward, 2012). Elimination of these errors must be carried out on a case-by-case basis, and is difficult and time consuming. For this reason, the use of specimen vouchers that are held in permanent collections and linked to reference sequences is preferable. The inclusion of multiple reference specimens representing each species can also help to prevent misidentification of unknown samples when queried against a reference library. Along these lines, the FDA has established a searchable database of standardized COI DNA barcode sequences called the reference standard sequence library (RSSL) for seafood identification (http:// www.accessdata.fda.gov/scripts/fdcc/?set=seafood_barcode_data). The specimens associated with these sequences have been taxonomically authenticated and, in most cases, are vouchered as part of permanent, curated collections at established institutions, such as the Smithsonian National Museum of Natural History. This database contains over 1000 sequence records representing a variety of vertebrate and invertebrate seafood species. The library is used by FDA to confirm labeling of seafood products sold commercially in the United States and details on its development are given in Deeds et al. (2014). One limitation of this dataset is that it lacks the broad geographic and taxonomic coverage that is needed to ensure that barcodes are diagnostic (Ekrem et al., 2007; Zhang et al., 2010). Adapting the potentially broad but error-prone public data sources for regulatory use represents an important area of ongoing research. Although formal approaches for their integration are currently lacking, querying an unknown sequence against multiple data resources is recommended.

An added benefit of the development of extensive DNA sequence reference libraries is the large quantity of sequence information that becomes available for the design of rapid DNA-based assays, such as species-specific PCR.

For example, in a series of studies focused on the differentiation of commercial salmon and trout species, a large collection of reference samples was sequenced across the DNA barcode region (Rasmussen et al., 2009) and the resulting sequence information was then used to develop a species-specific multiplex PCR assay for the rapid identification of salmonid species (Rasmussen Hellberg et al., 2010; Rasmussen Hellberg et al., 2011). Their potential portability and reduced cost make PCR-based assays attractive for high-throughput testing and they are discussed in more detail in the next chapter.

COMPUTATIONAL APPROACHES TO SEQUENCE-BASED SPECIES IDENTIFICATION

There are several methods in place that may be used to identify unknown samples to the species level once DNA sequence information has been obtained. As mentioned earlier, FINS and DNA barcoding rely on genetic distance calculations and phylogenetic analysis. Using these methods, an unknown sample can be identified to the species level by determining the genetic similarity/phylogenetic affinity of the query sequence with a set of reference sequences derived from expert-identified voucher specimens. One of the most commonly used methods for calculating genetic distance among seafood species is the Kimura 2-Parameter method (Kimura, 1980). In addition to evaluating the number of nucleotide substitutions between sequences, this method also considers the nature of the substitutions (i.e., transitions versus transversions) to determine evolutionary distance. This method is widely used in seafood identification studies based on both DNA barcoding (Hubert et al., 2008; Landi et al., 2014; Steinke et al., 2009) and FINS (Armani et al., 2015a; Chan et al., 2012; Huang et al., 2014). It is also a method available for use on BOLD when determining the genetic distances between a query sequence and reference sequences (http://www.boldsystems.org/index.php/IDS_OpenIdEngine). Alternatively, a number of FINS studies use the Tamura–Nei distance method (Ardura et al., 2010; Blanco et al., 2008; Chapela et al., 2002; Espiñeira et al., 2009; Lago et al., 2011; Santaclara et al., 2014), which is similar to the Kimura 2-Parameter method in that it considers transitions and transversions, but it also accounts for differences in nucleotide frequencies (Nei and Kumar, 2000; Tamura and Nei, 1993). The FDA protocol calls for the use of the less complex Jukes–Cantor method for calculating genetic distance (Handy et al., 2011b), which assumes an equal rate of nucleotide substitution and does not correct for transitional or transversional substitutions (Jukes and Cantor, 1969; Nei and Kumar, 2000), while others have argued that an uncorrected pairwise or "p" distance is suitable for assessing sequence similarity in barcode studies (Srivathsan and Meier, 2012).

Following the calculation of genetic distances, phylogenetic analysis is typically carried out with the neighbor-joining method (Saitou and Nei, 1987), which is able to handle large species assemblages and generate results relatively quickly (Hebert et al., 2003a; Kumar and Gadagkar, 2000). BOLD utilizes this method to display the relationship between a query sequence and the most closely related reference sequences (http://www.boldsystems.org/index.php/IDS_OpenIdEngine). BOLD also supports sequence clustering using an unpublished proprietary method based on a "refined single linkage" analysis (Ratnasingham and Hebert, 2013). Although this method presents a useful interim taxonomic framework for communicating about biodiversity at the genetic level, its adoption for regulatory identification purposes is currently without precedent. The FDA protocol calls for the use of either the neighbor-joining method or the Unweighted Pair Group Method with Arithmetic Mean (UPGMA) method, which is a simple method for construction of phylogenetic trees that allows for smoother branches when the sequence lengths in the dataset are variable (Handy et al., 2011b).

Another method for the identification of species based on DNA sequence information is called character-based analysis (Sarkar et al., 2008). Character-based analysis focuses on specific nucleotide sites known to vary between species rather than calculating genetic distance across all nucleotide sites in a given sequence. To identify a sample at the species level using character-based analysis, diagnostic sites must first be determined using a set of reference sequences. Character-based analysis may be used as a complement to the methods described earlier in circumstances that call for greater scrutiny, for example in regulatory or legal situations or when species are too closely related to be differentiated based on genetic distance alone (Wong et al., 2009). This method has been successfully used for species differentiation of selected groups of seafood, including tuna (Abdullah and Rehbein, 2014; Lowenstein et al., 2009; Richardson et al., 2006), sharks (Wong et al., 2009), freshwater fish of Cuba (Lara et al., 2010), and billfish (Richardson et al., 2006). However, it is important to keep in mind that the character-based approach is limited by the group of sequences and species originally used to determine diagnostic nucleotide sites. As the number of sequences representing each species grows, the complexity of the analysis will also increase due to the largely unsampled haplotype variation present within species, which appears to be quite high for fishes (Phillips et al., 2015). The analysis of diagnostic patterns to differentiate species will also become increasingly complicated as the species pool is expanded. A diversity of more computationally intensive approaches to sequence matching has been proposed (e.g., Zhang et al., 2008, 2011b), but because they are not commonly implemented using available barcode workflows, they are not discussed in depth here. However, it is important to realize that when an obvious species-level match cannot be obtained for a query sequence (e.g., due to incomplete reference libraries) other tools are available that can be used to place a sequence within a higher taxonomic framework. For example, Cohen et al. (2009) obtained a 100% posterior probability placement within the genus *Lagocephalus* for an unidentified fish sample's barcode sequence that had no species-level match in available databases at the time the analysis was undertaken.

NEXT-GENERATION SEQUENCING FOR USE IN SEAFOOD SPECIES IDENTIFICATION

Despite the numerous applications of species identification methods based on Sanger sequencing, these methods cannot readily be used to identify multiple species in a single sample (e.g., canned products, surimi) unless a time-consuming and expensive cloning step is added (Teletchea, 2009). Species-specific PCR methods are generally used in these instances; however, methods based on NGS are also capable of detecting multiple species simultaneously and have the potential to be used for testing mixed-species seafood products. In a related application, Tillmar et al. (2013) demonstrated the ability to use NGS to identify mammalian species in mixtures down to levels of 1% based on <100-bp regions of the 16S rRNA gene. NGS techniques amplify DNA fragments immobilized on a solid matrix through PCR to achieve high volumes of homologous strands of short read lengths (50–500 bp), a methodology also characterized as "massive parallel sequencing" (Zhang et al., 2011a). Simultaneous sequencing of millions of DNA fragments using NGS allows for retrieval of sequences of individual species in mixed-species samples (Shokralla et al., 2014). Due to the limited sequence read lengths acquired through NGS, the utilization of short DNA fragments or mini-barcodes is preferable for seafood species identification applications. For example, De Battisti et al. (2014) reported the ability to differentiate 25 species of fish from the *Clupeiformes* and *Pleuronectiformes* groups using NGS of 30-bp regions of 16S rRNA, cyt *b*, and the NADH dehydrogenase subunit II genes. Some promising NGS platforms currently available that show high potential for use in seafood species identification include MiSeq (Illumina) and Ion Proton by Ion Torrent (Life Technologies).

The MiSeq system is based on the principle of "sequencing by synthesis" (SBS), in which fluorescently labeled reversible deoxynucleotide (dNTP) terminators attach to a single-stranded DNA template (Illumina, 2015). As each dNTP is added to the growing DNA chain, a florescent signal of characteristic wavelength and intensity is emitted identifying each base. According to the manufacturer, the MiSeq platform has a maximum sequence read length of 300 bp and can handle up to 15 GB of data. Run times for the instrument range from 5 to 55 h, depending on the read length and total data output desired (Illumina, 2015). In an example of fish identification applications, this platform has been utilized to census bony fish to the genus or family level based on analysis of a 106-bp fragment of the 12S rRNA gene recovered from environmental DNA (Kelly et al., 2014). Besides the potential application of NGS techniques to identify seafood species in mixtures, the large amount of sequence data generated can be used to identify genetic markers for rapid assays. For example, the MiSeq platform was used to perform sequencing of the mitochondrial genomes of two invasive carp species (*Hypopthalmichthys nobilis* and *Hypopthalmichthys molitrix*), resulting in the development of rapid, species-specific PCR assays for use in routine surveillance of these organisms (Farrington et al., 2015).

In contrast to MiSeq, Ion Torrent technology monitors pH changes during dNTP incorporation into nascent stands by DNA polymerase (Life Technologies, 2015). With this method, only one type of nucleotide (A, G, C, or T) is added to the reaction at a time. When a nucleotide is incorporated into a complementary DNA strand, a hydrogen ion is released, causing a drop in pH that is used to monitor base addition. According to the manufacturer, reliable sequence lengths of 200–400 bp at a total output of up to 2 GB can be generated in run times of 2–4 h (Life Technologies, 2015). The Ion Torrent platform has been used to successfully identify mixtures of four fish species in the scat of harbor seals (*Phoca vitulina*) based on a ~120-bp region of the 16S rRNA gene (Deagle et al., 2013). This platform has also been used to study microsatellites of the fish species Marinka (*Schizothrox biddulphi*; Luo et al., 2012) and Sichuan Taiman (*Hucho bleekeri*; Wang et al., 2015).

Despite the power of MiSeq and Ion Torrent technologies, each has specific shortcomings (Glenn, 2011; Loman et al., 2012). For instance, the MiSeq platform is more expensive than the Ion Torrent platform, both overall and in terms of cost per run, and it demands significantly longer run times (for current specifications, see http://www.molecularecologist.com/next-gen-fieldguide-2014/). On the other hand, Ion Torrent technology is more labor intensive, requiring emulsion PCR, has a lower throughput, and produces shorter read lengths compared to MiSeq (Loman et al., 2012) which may not yield sufficient information for accurate species matching. It is important to note that NGS technology is a rapidly growing area of research and many of these issues will likely be improved upon in the coming years. Furthermore, numerous NGS platforms are currently in various stages of development and may ultimately supplant those mentioned here. Unlike Sanger sequencing, none of the NGS platforms are currently recognized as being appropriate for diagnostic applications—they are research instruments. Moreover, the use of NGS platforms is complicated by the lack of accessible analytical pipelines capable of handling the large volumes of data they produce. Yet, as this technology becomes more accessible, it is also likely that it will be increasingly used for applications in seafood species identification.

APPLICATION OF DNA SEQUENCING TO DETECT COMMERCIAL SEAFOOD FRAUD

A major application of the DNA sequencing methods described earlier has been the detection of seafood mislabeling on the commercial market. As shown in Table 6.1, studies of this nature have been carried out globally and have revealed widespread mislabeling of some commercial seafood products. These studies have utilized the DNA barcoding and FINS methodologies discussed earlier to differentiate between species based on genetic targets such as COI, 16S rRNA, and cyt *b*. Studies of mislabeled commercial seafood products have revealed that the most common incentive for species substitution is economic, with the

TABLE 6.1 Examples of Studies That Have Utilized DNA Sequencing to Detect Seafood Mislabeling on the Commercial Market

Seafood Species Targeted	Product Type	Geographic Location	Species Identification Method	Target Gene	Study Findings	References
Clupeidae, Engraulidae, Salangidae, and Scombridae families	Canned cat food	Italy	FINS and BLAST	16S rRNA	Mislabeling rates of 40–100%, depending on the product category	Armani et al. (2015b)
Jellyfish (Class Scyphozoa)	Dry salted and ready-to-eat products	Italy	FINS	COI	79% of dry salted and 100% of ready-to-eat products were mislabeled	Armani et al. (2013)
Abalone (Haliotidae family)	Dried, canned abalone	Hong Kong	FINS and BLAST	16S rRNA	One canned sample and all sliced (dried) samples were mislabeled	Chan et al. (2012)
Puffer fish (Lagocephalus and Takifugu)	Fillets, canned products and fish powders	Taiwan	FINS and BLAST	cyt b	14% of samples had puffer fish, but 28% of those were toxic varieties, causing a potential threat	Huang et al. (2014)
Freshwater fishes (Cyprinidae family)	Whole fish	Taiwan	FINS and BLAST	cyt b	38% of commercial samples did not contain Cyprinidae	Chen et al. (2012)
Atlantic salmon (Salmo salar) and Pacific salmon (Oncorhynchus spp.)	Fillets	United States	DNA barcoding	COI	11% of samples labeled as Pacific salmon were Atlantic salmon and restaurant substitution was more common than supermarket substitution	Cline (2012)
Gadidae family	Salted fillets and battered chunks	Italy	DNA barcoding	COI	85% of salted fillets were substituted with Lotidae species and 100% of battered cod chunks were substituted with pollack (Pollachius virens) and tusk (Brosme brosme)	Di Pinto et al. (2013)

Species	Product	Country	Method	Gene	Findings	Reference
Sciaenidae family (*Plagioscion squamosissimus* and *Cynoscion leiarchus*)	Fillets	Brazil	DNA barcoding	COI	Mislabeling rates of 77–100% depending on labeling criteria	De Brito et al. (2015)
Tilapia (*Oreochromis* spp.), Nile perch (*Lates niloticus*), and panga (*Pangasius* or *Pangasionodon* spp.)	Fillets	Egypt	DNA barcoding	COI	50% of Nile perch and panga samples were substituted with tra fish (*Pangasianodon hypophthalmus*) and none of the tilapia was mislabeled	Galal-Khallaf et al. (2014)
Variety of commonly consumed species	Fillets	United States	DNA barcoding	COI	16.8% of samples collected from restaurants were mislabeled	Khaksar et al. (2015)
	Fillets	Tasmania	DNA barcoding	COI	No samples in the study were mislabeled	Lamendin et al. (2015)
European plaice (*Pleuronectes platessa*) and common sole (*Solea solea*)	Fillets	Italy	DNA barcoding	COI	41% of common sole samples and 35% of European plaice samples were mislabeled	Pappalardo and Ferrito (2015)

BLAST, Basic Local Alignment Search Tool; COI, cytochrome c oxidase subunit I; cyt b, cytochrome b; FINS, forensically informative nucleotide sequencing.

substituted species being less expensive than the labeled variety. For example, tra fish (*Pangasianodon hypophthalmus*) commonly substitutes for the more expensive Nile perch (*Lates niloticus*; Galal-Khallaf et al., 2014); Atlantic salmon (*Salmo salar*) for Pacific salmon (*Oncorhynchus* spp.; Cline, 2012); and tusk (*Brosme brosme*) for cod (*Gadus* spp.; Di Pinto et al., 2013). In one representative study, Di Pinto et al. (2013) used COI DNA barcoding to examine the prevalence of commercial salted cod and battered cod products in Italy, where, by law all commercial products labeled as "cod" must be from *Gadus macrocephalus* or *Gadus morhua*. The authors found that 85% of salted and 100% of battered cod products contained the less expensive tusk fish (*B. brosme*).

A less insidious substitution trend is accidental substitution, when species that bare morphological similarities and occupy common habitats are harvested together. In a COI DNA barcoding study conducted by Pappalardo and Ferrito (2015), European flounder (*Platichthys flesus*), common dab (*Limanda limanda*), and iridescent shark (*Pangasius hypophthalmus*) were often found substituted for the morphologically similar European plaice (*Pleuronectes platessa*), an adulteration that, at least in some cases, was presumed accidental. One of the most disturbing adulteration trends is the substitution of toxic varieties of seafood for nontoxic species. This may involve the presence of specific toxins posing broad risk or allergens to which a segment of the population is vulnerable. For example, through FINS and BLAST analysis of cyt *b*, Huang et al. (2014) found that 28% of commercially labeled puffer fish (*Lagocephalus gloveri* and *Lagocephalus wheeleri*) were substituted with more toxic varieties of puffer fish from the genus *Takifugu*. Overall, DNA sequencing has proven to be extremely useful in the detection of seafood fraud on the commercial market. Through applications such as those described earlier, it can be used as a tool to assist in preventing incidences of public health and battling economic fraud of seafood products.

CONCLUSIONS

In conclusion, Sanger DNA sequencing is an essential and well-established method for the identification of seafood species. To use this method for the reliable identification of species, an appropriate genetic target and reference database must be selected. In addition to its widespread use in the differentiation of species, this method also results in the generation of sequence data that can be used for the development of rapid, targeted assays for species detection. The standardization of DNA sequencing protocols through the DNA barcoding movement has enabled widespread use of COI as a genetic marker for seafood species identification as well as the development of extensive reference databases. More effort is required to expand geographic coverage within species as well as broad taxonomic coverage between them to address seafood fraud on a global scale, but most commercially exploited species already have some barcode coverage. Similarly, more work is needed to define appropriate nDNA targets and assess the potential for hybridization, but this issue does not concern many detected cases of substitution that are economically motived and do not involve closely related species. Finally, advances in techniques such as mini-barcoding and NGS will enable DNA sequencing to be readily applied to the analysis of heavily processed and mixed-species seafood products.

REFERENCES

Abdullah, A., Rehbein, H., 2014. Authentication of raw and processed tuna from Indonesian markets using DNA barcoding, nuclear gene and character-based approach. European Food Research & Technology 239, 695–706.

Ardura, A., Pola, I.G., Linde, A.R., Garcia-Vazquez, E., 2010. DNA-based methods for species authentication of Amazonian commercial fish. Food Research International 43, 2295–2302.

Armani, A., Guardone, L., Castigliego, L., D'Amico, P., Messina, A., Malandra, R., Gianfaldoni, D., Guidi, A., 2015a. DNA and Mini-DNA barcoding for the identification of Porgies species (family Sparidae) of commercial interest on the international market. Food Control 50, 589–596.

Armani, A., Tinacci, L., Giusti, A., Castigliego, L., Gianfaldoni, D., Guidi, A., 2013. What is inside the jar? Forensically informative nucleotide sequencing (FINS) of a short mitochondrial COI gene fragment reveals a high percentage of mislabeling in jellyfish food products. Food Research International 54, 1383–1393.

Armani, A., Tinacci, L., Xiong, X., Castigliego, L., Gianfaldoni, D., Guidi, A., 2015b. Fish species identification in canned pet food by BLAST and Forensically Informative Nucleotide Sequencing (FINS) analysis of short fragments of the mitochondrial 16s ribosomal RNA gene (16S rRNA). Food Control 50, 821–830.

Bhattacharjee, M.J., Ghosh, S.K., 2014. Design of mini-barcode for catfishes for assessment of archival biodiversity. Molecular Ecology Resources 14, 469–477.

Blanco, M., Perez-Martin, R.I., Sotelo, C.G., 2008. Identification of shark species in seafood products by forensically informative nucleotide sequencing (FINS). Journal of Agricultural and Food Chemistry 56, 9868–9874.

Borisenko, A.V., Sones, J.E., Hebert, P.D.N., 2009. The front-end logistics of DNA barcoding: challenges and prospects. Molecular Ecology Resources 9, 27–34. http://dx.doi.org/10.1111/j.1755-0998.2009.02629.x.

Calo-Mata, P., Sotelo, C.G., Perez-Martin, R.I., Rehbein, H., Hold, G.L., Russell, V.J., Pryde, S., Quinteiro, J., Rey-Mendez, M., Rosa, C., Santos, A.T., 2003. Identification of gadoid fish species using DNA-based techniques. European Food Research & Technology 217, 259–264.

Carvalho, D.C., Palhares, R.M., Drummond, M.G., Frigo, T.B., 2015. DNA barcoding identification of commercialized seafood in South Brazil: a governmental regulatory forensic program. Food Control 50, 784–788.

Cawthorn, D.-M., Steinman, H.A., Witthuhn, R.C., 2011. Establishment of a mitochondrial DNA sequence database for the identification of fish species commercially available in South Africa. Molecular Ecology Resources 11, 979–991.

CBOL, 2015. Consortium for the Barcode of Life. Barcode of Life: Identifying Species with DNA Barcoding. Accessible at: http://www.barcodeoflife.org/content/about/what-dna-barcoding.

Chan, W.-H., Ling, K.-H., Shaw, P.-C., Chiu, S.-W., Pui-Hay But, P., 2012. Application of FINS and multiplex PCR for detecting genuine abalone products. Food Control 23, 137–142.

Chapela, M.J., Sotelo, C.G., Calo-Mata, P., Perez-Martin, R.I., Rehbein, H., Hold, G.L., Quinteiro, J., Rey-Mendez, M., Rosa, C., Santos, A.T., 2002. Identification of cephalopod species (Ommastrephidae and Loliginidae) in seafood products by forensically informative nucleotide sequencing (FINS). Journal of Food Science 67, 1672–1676.

Chen, C.-H., Hsieh, C.-H., Hwang, D.-F., 2012. Species identification of Cyprinidae fish in Taiwan by FINS and PCR–RFLP analysis. Food Control 28, 240–245.

Chow, S., Nakagawa, T., Suzuki, N., Takeyama, H., Matsunaga, T., 2006. Phylogenetic relationships among *Thunnus* species inferred from rDNA ITS1 sequence. Journal of Fish Biology 68, 24–35.

Cline, E., 2012. Marketplace substitution of Atlantic salmon for Pacific salmon in Washington State detected by DNA barcoding. Food Research International 45, 388–393.

Cohen, N.J., Deeds, J.R., Wong, E.S., Hanner, R.H., Yancy, H.F., White, K.D., Thompson, T.M., Wahl, M., Pham, T.D., Guichard, F.M., Huh, I., Austin, C., Dizikes, G., Gerber, S.I., 2009. Public health response to puffer fish (Tetrodotoxin) poisoning from mislabeled product. Journal of Food Protection 72, 810–817.

Collins, R.A., Armstrong, K.F., Meier, R., Yi, Y., Brown, S.D.J., Cruickshank, R.H., et al., 2012. Barcoding and border biosecurity: identifying cyprinid fishes in the aquarium trade. PLoS One 7 (1), e28381. http://dx.doi.org/10.1371/journal.pone.0028381.

Cutarelli, A., Amoroso, M.G., De Roma, A., Girardi, S., Galiero, G., Guarino, A., Corrado, F., 2014. Italian market fish species identification and commercial frauds revealing by DNA sequencing. Food Control 37, 46–50.

Dalmasso, A., Chiesa, F., Civera, T., Bottero, M.T., 2013. A novel minisequencing test for species identification of salted and dried products derived from species belonging to Gadiformes. Food Control 34, 296–299.

De Battisti, C., Marciano, S., Magnabosco, C., Busato, S., Arcangeli, G., Cattoli, G., 2014. Pyrosequencing as a tool for rapid fish species identification and commercial fraud detection. Journal of Agricultural and Food Chemistry 62, 198–205.

De Brito, M.A., Schneider, H., Sampaio, I., Santos, S., 2015. DNA barcoding reveals high substitution rate and mislabeling in croaker fillets (Sciaenidae) marketed in Brazil: the case of 'pescada branca' (*Cynoscion leiarchus* and *Plagioscion squamosissimus*). Food Research International 70, 40–46.

Deagle, B.E., Thomas, A.C., Shaffer, A.K., Trites, A.W., Jarman, S.N., 2013. Quantifying sequence proportions in a DNA-based diet study using Ion Torrent amplicon sequencing: which counts count? Molecular Ecology Resources 13, 620–633.

Deeds, J.R., Handy, S.M., Fry Jr., F., Granade, H., Williams, J.T., Powers, M., Shipp, R., Weigt, L.A., 2014. Protocol for building a reference standard sequence library for DNA-based seafood identification. Journal of AOAC International 1626–1633.

Di Pinto, A., Di Pinto, P., Terio, V., Bozzo, G., Bonerba, E., Ceci, E., Tantillo, G., 2013. DNA barcoding for detecting market substitution in salted cod fillets and battered cod chunks. Food Chemistry 141, 1757–1762.

Ekrem, T., Willassen, E., Stur, E., 2007. A comprehensive DNA sequence library is essential for identification with DNA barcodes. Molecular Phylogenetics and Evolution 43, 530–542.

Espiñeira, M., Gonzalez-Lavin, N., Vieites, J.M., Santaclara, F.J., 2009. Development of a method for the identification of scombroid and common substitute species in seafood products by FINS. Food Chemistry 117, 698–704.

Evans, N., Paulay, G., 2012. DNA barcoding methods for invertebrates. In: Kress, W.J., Erickson, D.L. (Eds.), DNA Barcodes: Methods and Protocols. Springer, New York, pp. 47–77.

Farrington, H.L., Edwards, C.E., Guan, X., Carr, M.R., Baerwaldt, K., Lance, R.F., 2015. Mitochondrial genome sequencing and development of genetic markers for the detection of DNA of invasive bighead and silver carp (*Hypophthalmichthys nobilis* and *H. molitrix*) in environmental water samples from the United States. PLoS One 10, 1–17.

Galal-Khallaf, A., Ardura, A., Mohammed-Geba, K., Borrell, Y.J., Garcia-Vazquez, E., 2014. DNA barcoding reveals a high level of mislabeling in Egyptian fish fillets. Food Control 46, 441–445.

Glenn, T.C., 2011. Field guide to next generation DNA sequencers. Molecular Ecology Resources 11, 759–769.

Hajibabaei, M., McKenna, C., 2012. DNA mini-barcodes. In: Kress, W.J., Erickson, D.L. (Eds.), DNA Barcodes: Methods and Protocols. Springer, New York, pp. 339–353.

Hajibabaei, M., Smith, M.A., Janzen, D.H., Rodriguez, J.J., Whitfield, J.B., Hebert, P.D.N., 2006. A minimalist barcode can identify a specimen whose DNA is degraded. Molecular Ecology Notes 6, 959–964.

Handy, S., Deeds, J., Ivanova, N., Hebert, P., Hanner, R., Ormos, A., Weigt, L., Moore, M., Yancy, H., 2011a. A single laboratory validated method for the generation of DNA barcodes for the identification of fish for regulatory compliance. Journal of AOAC International 94, 201–210.

Handy, S.M., Deeds, J.R., Ivanova, N.V., Hebert, P.D.N., Hanner, R.H., Ormos, A., Weigt, L.A., Moore, M.M., Hellberg, R.S., Yancy, H.F., 2011b. Single Laboratory Validated Method for DNA-Barcoding for the Species Identification of Fish for FDA Regulatory Compliance. U.S. Food and Drug Administration Standard Operating Procedure. Updated September 2011. Accessible at: http://www.fda.gov/food/foodscienceresearch/dnaseafoodidentification/ucm237391.htm.

Harris, D.J., 2003. Can you bank on GenBank? Trends in Ecology and Evolution 18, 317–319.

Hebert, P.D.N., Cywinska, A., Ball, S.L., deWaard, J.R., 2003a. Biological identifications through DNA barcodes. Proceedings of the Royal Society of London. Series B: Biological Sciences 270, 313–321.

Hebert, P.D.N., Ratnasingham, S., deWaard, J.R., 2003b. Barcoding animal life: cytochrome *c* oxidase subunit 1 divergences among closely related species. Proceedings of the Royal Society of London. Series B: Biological Sciences 270, S96–S99.

Huang, Y.-R., Yin, M.-C., Hsieh, Y.-L., Yeh, Y.-H., Yang, Y.-C., Chung, Y.-L., Hsieh, C.-H.E., 2014. Authentication of consumer fraud in Taiwanese fish products by molecular trace evidence and forensically informative nucleotide sequencing. Food Research International 55, 294–302.

Hubert, N., Hanner, R., Holm, E., Mandrak, N.E., Taylor, E., Burridge, M., Watkinson, D., Dumont, P., Curry, A., Bentzen, P., Zhang, J., April, J., Bernatchez, L., 2008. Identifying Canadian freshwater fishes through DNA barcodes. PLoS One 3, e2490.

Illumina, 2015. An Introduction to Next-Generation Sequencing. Illumina, Inc., San Diego, CA. Available from: http://www.illumina.com/content/dam/illumina-marketing/documents/products/illumina_sequencing_introduction.pdf.

Jukes, T.H., Cantor, C.R., 1969. Evolution of protein molecules. In: Munro, H.N. (Ed.), Mammalian Protein Metabolism. Academic Press, New York, pp. 21–132.

Kelly, R.P., Port, J.A., Yamahara, K.M., Crowder, L.B., 2014. Using environmental DNA to census marine fishes in a large mesocosm. PLoS One 9, 1–11.

Khaksar, R., Carlson, T., Schaffner, D.W., Ghorashi, M., Best, D., Jandhyala, S., Traverso, J., Amini, S., 2015. Unmasking seafood mislabeling in U.S. markets: DNA barcoding as a unique technology for food authentication and quality control. Food Control 56, 71–76.

Kim, D.-W., Yoo, W.G., Park, H.C., Yoo, H.S., Kang, D.W., Jin, S.D., Min, H.K., Paek, W.K., Lim, J., 2012. DNA barcoding of fish, insects, and shellfish in Korea. Genomics & Informatics 10, 206–211.

Kimura, M., 1980. A simple method of estimating evolutionary rate of base substitutions through comparative studies of nucleotide sequences. Journal of Molecular Evolution 16, 111–120.

Kocher, T.D., Thomas, W.K., Meyer, A., Edwards, S.V., Pääbo, S., Villablanca, F.X., Wilson, A.C., 1989. Dynamics of mitochondrial DNA evolution in animals: amplification and sequencing with conserved primers. Proceedings of the National Academy of Sciences of the United States of America 86, 6196–6200.

Kochzius, M., Seidel, C., Antoniou, A., Botla, S.K., Campo, D., Cariani, A., Vazquez, E.G., Hauschild, J., Hervet, C., Hjörleifsdottir, S., Hreggvidsson, G., Kappel, K., Landi, M., Magoulas, A., Marteinsson, V., Nölte, M., Planes, S., Tinti, F., Turan, C., Venugopal, M.N., Weber, H., Blohm, D., 2010. Identifying fishes through DNA barcodes and microarrays. PLoS One 5, e12620.

Kress, W.J., Erickson, D.L. (Eds.), 2012. DNA Barcodes: Methods and Protocols. Methods in Molecular Biology, Humana Press, New Jersey, pp. 17–46.

Kumar, S., Gadagkar, S.R., 2000. Efficiency of the neighbour-joining method in reconstructing deep and shallow evolutionary relationships in large phylogenies. Journal of Molecular Evolution 51, 544–553.

Lago, F.C., Herrero, B., Vieites, J.M., Espiñeira, M., 2011. Genetic identification of horse mackerel and related species in seafood products by means of forensically informative nucleotide sequencing methodology. Journal of Agricultural and Food Chemistry 59, 2223–2228.

Lamendin, R., Miller, K., Ward, R.D., 2015. Labelling accuracy in Tasmanian seafood: an investigation using DNA barcoding. Food Control 47, 436–443.

Landi, M., Dimech, M., Arculeo, M., Biondo, G., Martins, R., Carneiro, M., Carvalho, G.R., Lo Brutto, S., Costa, F.O., 2014. DNA barcoding for species assignment: the case of Mediterranean marine fishes. PLoS One 9, e106135.

Lara, A., Ponce de León, J.L., Rodríguez, R., Casane, D., Côtes, G., Bernatchez, L., García-Machado, E., 2010. DNA barcoding of Cuban freshwater fishes: evidence for cryptic species and taxonomic conflicts. Molecular Ecology Resources 10, 421–430.

Life Technologies, 2015. Ion PGM System Specifications. Thermo Fisher Scientific Inc., Carlsbad, CA. Available from: http://www.lifetechnologies.com/us/en/home/brands/ion-torrent.html.

Liu, J.U.N., Li, Q.I., Kong, L., Yu, H., Zheng, X., 2011. Identifying the true oysters (Bivalvia: Ostreidae) with mitochondrial phylogeny and distance-based DNA barcoding. Molecular Ecology Resources 11, 820–830.

Loman, N.J., Misra, R.V., Dallman, T.J., Constantinidou, C., Gharbia, S.E., Wain, J., Pallen, M.J., 2012. Performance comparison of benchtop high-throughput sequencing platforms. Nature Biotechnology 30, 434–439.

Lowenstein, J.H., Amato, G., Kolokotronis, S.O., 2009. The real maccoyii: identifying tuna sushi with DNA barcodes – contrasting characteristic attributes and genetic distances. PLoS One 4, e7866.

Luo, W., Nie, Z.L., Zhan, F.B., Wei, J., Wang, W.M., Gao, Z.X., 2012. Rapid development of microsatellite markers for the endangered fish *Schizothorax biddulphi* (Gunther) using next generation sequencing and cross-species amplification. International Journal of Molecular Sciences 13, 14946–14955.

Meusnier, I., Singer, G.A.C., Landry, J.F., Hickey, D.A., Hebert, P.D.N., Hajibabaei, M., 2008. A universal DNA mini-barcode for biodiversity analysis. BMC Genomics 9, 1–4.

Nei, M., Kumar, S., 2000. Molecular Evolution and Phylogenetics. Oxford University Press, New York.

Pappalardo, A.M., Ferrito, V., 2015. DNA barcoding species identification unveils mislabeling of processed flatfish products in southern Italy markets. Fisheries Research 164, 153–158.

Phillips, J.D., Gwiazdowski, R.A., Ashlock, D., Hanner, R., 2015. An exploration of sufficient sampling effort to describe intraspecific DNA barcode haplotype diversity: examples from the ray-finned fishes (Chordata: Actinopterygii). DNA Barcodes 3, 66–73.

Rasmussen Hellberg, R.S., Morrissey, M.T., 2011. Advances in DNA-based techniques for the detection of seafood species substitution on the commercial market. Journal of Laboratory Automation 16, 308–321.

Rasmussen Hellberg, R.S., Morrissey, M.T., Hanner, R.H., 2010. A multiplex PCR method for the identification of commercially important salmon and trout species (*Oncorhynchus* and *Salmo*) in North America. Journal of Food Science 75, 595–606.

Rasmussen Hellberg, R.S., Naaum, A.M., Handy, S.M., Hanner, R.H., Deeds, J.R., Yancy, H.F., Morrissey, M.T., 2011. Interlaboratory evaluation of a real-time multiplex polymerase chain reaction method for identification of salmon and trout species in commercial products. Journal of Agricultural and Food Chemistry 59, 876–884.

Rasmussen, R.S., Morrissey, M.T., 2008. DNA-based methods for the identification of commercial fish and seafood species. Comprehensive Reviews in Food Science and Food Safety 7, 280–295.

Rasmussen, R.S., Morrissey, M.T., 2009. Application of DNA-based methods to identify fish and seafood substitution on the commercial market. Comprehensive Reviews in Food Science and Food Safety 8, 118–154.

Rasmussen, R.S., Morrissey, M.T., Hebert, P.D.N., 2009. DNA barcoding of commercially important salmon and trout species (*Oncorhynchus* and *Salmo*) from North America. Journal of Agricultural and Food Chemistry 57, 8379–8385.

Ratnasingham, S., Hebert, P.D.N., 2013. A DNA-Based registry for all animal species: the Barcode Index Number (BIN) System. PLoS One 8 (7), e66213.

Rehbein, H., 2013. Differentiation of fish species by PCR-based DNA analysis of nuclear genes. European Food Research & Technology 236, 979–990.

Richardson, D.E., Vanwy, J.D., Exum, A.M., Cowen, R.K., Crawford, D.L., 2006. High-throughput species identification: from DNA isolation to bioinformatics. Molecular Ecology Notes 7, 199–207.

Saitou, N., Nei, M., 1987. The neighbor-joining method: a new method for reconstructing phylogenetic trees. Molecular Biology and Evolution 4, 406–425.

Sanger, F., Nicklen, S., Coulson, A.R., 1977. DNA sequencing with chain-terminating inhibitors. Proceedings of the National Academy of Sciences 74, 5463–5467.

Santaclara, F.J., Perez-Martin, R.I., Sotelo, C.G., 2014. Developed of a method for the genetic identification of ling species (*Genypterus* spp.) in seafood products by FINS methodology. Food Chemistry 143, 22–26.

Sarkar, I.N., Planet, P.J., DeSalle, R., 2008. CAOS software for use in character-based DNA barcoding. Molecular Ecology Resources 8, 1256–1259.

Sevilla, R.G., Diez, A., Norén, M., Mouchel, O., Jérôme, M., Verrez-Bagnis, V., Van Pelt, H., Favre-Krey, L., Krey, G., Consortium, T.F., Bautista, J.M., 2007. Primers and polymerase chain reaction conditions for DNA barcoding teleost fish based on the mitochondrial cytochrome *b* and nuclear rhodopsin genes. Molecular Ecology Notes 7, 730–734.

Shokralla, S., Gibson, J.F., Nikbakht, H., Janzen, D.H., Hallwachs, W., Hajibabaei, M., 2014. Next-generation DNA barcoding: using next-generation sequencing to enhance and accelerate DNA barcode capture from single specimens. Molecular Ecology Resources 14, 892–901.

Shokralla, S., Hellberg, R.S., Handy, S.M., King, I., Hajibabaei, M., 2015. A DNA mini-barcoding system for authentication of processed fish products. Scientific Reports 5 Article number 15894.

Srivathsan, A., Meier, R., 2012. On the inappropriate use of Kimura-2-parameter (K2P) divergences in the DNA-barcoding literature. Cladistics 28, 190–194.

Steinke, D., Hanner, R., 2011. The FISH-BOL collaborators' protocol. Mitochondrial DNA 22 (S1), 10–14.

Steinke, D., Zemlak, T.S., Gavin, H., Hebert, P.D.N., 2009. DNA barcoding of Pacific Canada's fishes. Marine Biology 156, 2641–2647.

Tamura, K., Nei, M., 1993. Estimation of the number of nucleotide substitutions in the control region of mitochondrial DNA in humans and chimpanzees. Molecular Biology and Evolution 10, 512–526.

Teletchea, F., 2009. Molecular identification methods of fish species: reassessment and possible applications. Reviews in Fish Biology and Fisheries 19, 265–293.

Tillmar, A.O., Dell'Amico, B., Welander, J., Holmlund, G., 2013. A universal method for species identification of mammals utilizing next generation sequencing for the analysis of DNA mixtures. PLoS One 8, e83761.

Unseld, M., Beyermann, B., Brandt, P., Hiesel, R., 1995. Identification of the species origin of highly processed meat products by mitochondrial DNA sequences. PCR Methods and Applications 4, 241–243.

Viñas, J., Tudela, S., 2009. A validated methodology for genetic identification of tuna species (Genus *Thunnus*). PLoS One 4, e7606.

Wang, K., Zhang, S., Wang, D., Xin, M., Wu, J., Sun, Q., Du, H., Wang, C., Huang, J., Wei, Q., 2015. Development of 27 novel cross-species microsatellite markers for the endangered *Hucho bleekeri* using next-generation sequencing technology. Conservation Genetics Resources 7, 263–267.

Ward, R.D., 2012. FISH-BOL, a case study for DNA barcodes. In: Kress, W.J., Erickson, D.L. (Eds.), DNA Barcodes: Methods and Protocols. Springer, New York, pp. 423–439.

Ward, R.D., Hanner, R., Hebert, P.D.N., 2009. The campaign to DNA barcode all fishes, FISH-BOL. Journal of Fish Biology 74, 329–356.

Ward, R.D., Zemlak, T.S., Innes, B.H., Last, P.R., Hebert, P.D.N., 2005. DNA barcoding Australia's fish species. Philosophical Transactions of the Royal Society B: Biological Sciences 360, 1847–1857.

Weigt, L.A., Driskell, A.C., Baldwin, C.C., Ormos, A., 2012. DNA barcoding fishes. In: Kress, W.J., Erickson, D.L. (Eds.), DNA Barcodes: Methods and Protocols. Springer, New York, pp. 109–126.

Wong, E.H.K., Shivji, M.S., Hanner, R.H., 2009. Identifying sharks with DNA barcodes: assessing the utility of a nucleotide diagnostic approach. Molecular Ecology Resources 9, 243–256.

Yancy, H.F., Zemlak, T.S., Mason, J.A., Washington, J.D., Tenge, B.J., Nguyen, N.T., Barnett, J.D., Savary, W.E., Hill, W.E., Moore, M.M., Fry, F.S., Randolph, S.C., Rogers, P.L., Hebert, P.D.N., 2008. Potential use of DNA barcodes in regulatory science: applications of the regulatory fish encyclopedia. Journal of Food Protection 71, 456–458.

Zhang, J., Chiodini, R., Badr, A., Zhang, G., 2011a. The impact of next-generation sequencing on genomics. Journal of Genetics and Genomics 38, 95–109.

Zhang, J., Hanner, R., 2012. Molecular approach to the identification of fish in the South China Sea. PLoS One 7, e30621.

Zhang, A.-B., He, L.J., Crozier, R.H., Muster, C., Zhu, C.-D., 2010. Estimating sample sizes for DNA barcoding. Molecular Phylogenetics and Evolution 54 (3), 1035–1039.

Zhang, A.-B., Muster, C., Liang, H.-B., Zhu, C.-D., Crozier, R., Wan, P., et al., 2011b. A fuzzy-set-theory-based approach to analyse species membership in DNA barcoding. Molecular Ecology 21 (8), 1848–1863.

Zhang, A.-B., Sikes, D.S., Muster, C., Li, S.Q., 2008. Inferring species membership using DNA sequences with back-propagation neural networks. Systematic Biology 57 (2), 202–215.

Chapter 7

Species Identification Using Other Tools

LeeAnn Applewhite[1], Patrick Larkin[1], Amanda M. Naaum[2]

[1]Applied Food Technologies, Inc., Alachua, FL, United States; [2]University of Guelph, Guelph, ON, Canada

As mentioned in the previous chapter, sequencing is the most reliable and comprehensive way to obtain information from PCR fragments. However, it is a time-consuming and expensive process, making it impractical for routine use in many laboratories (Applewhite et al., 2012). Additionally, sequencing is not appropriate for the analysis of samples containing multiple species (Lenstra, 2003). The universal gene targets that are predominately used in sequencing often do not effectively discriminate some popular fish species such as tuna of the *Thunnus* genus (Pardo and Perez-Villareal, 2004; Lin and Hwang, 2007), and snappers such as American red snapper and Caribbean red snapper (Bayha et al., 2008). In processed seafood, DNA suffers degradation not only due to the thermal treatments that food muscle tissue is subjected to during the canning process (cooking and sterilization), but also to the type and pH of liquid (brine, oil, vinegar, or tomato) added during this processing (Bauer et al., 2003). Thus, several interrelated problems arise: (1) it may not be possible to extract high-quality DNA for sequencing, (2) the extracted DNA is not sufficient in size to be amplified for further analyses, and (3) the DNA from processed foods is often contaminated with substances that inhibit the PCR and amplification. Lastly, many traditional sequencing techniques cannot identify individual species in seafood items containing multiple species (Applewhite et al., 2012). To address these limitations with DNA sequencing, other molecular methods are often utilized in seafood species identification.

The most popular alternative methods to PCR sequencing in seafood species identification testing are species-specific PCR including multiplex PCR and real-time PCR, and restriction fragment length polymorphism (PCR RFLP). These methods are discussed first, followed by additional PCR-based methods that have been used less frequently for seafood species identification. Lastly, several emerging and promising technologies are presented.

Seafood Authenticity and Traceability. http://dx.doi.org/10.1016/B978-0-12-801592-6.00007-3
133

SPECIES-SPECIFIC PRIMERS, MULTIPLEX PCR, AND REAL-TIME PCR

Species-specific PCR has been widely used for years to identify various seafood species. Essential in species-specific PCR, the DNA sequence of the target species must be studied and compared within a species, as well as across closely related species, to determine nucleotide sites that differentiate the target species from other species. Once identified, species-specific primers are designed to bind and amplify the DNA at the discriminating nucleotide sites. PCR amplification is based on the hybridization of these oligonucleotides (primers) and synthesis of millions of copies of genetic material bordered by those primers (Mafra et al., 2007).

Traditional PCR

Traditionally, species-specific primers are designed to amplify a DNA fragment of a certain size. This can be done for a single target species at a time, or as a multiplex PCR. In multiplex PCR, multiple species are analyzed in a single reaction by using a combination of species-specific primers and/or universal primers, resulting in DNA fragment sizes that vary for the different species tested (Apte and Daniel, 2003). If the complete amplicon sequence is known for a target species, the size of the fragments can be predicted and that species can be identified by the appearance of an amplicon of appropriate size with agarose gel electrophoresis. Multiplex PCR with both mitochondrial and nuclear genes has been used for species diagnosis of a variety of species such as shark (Mendonc et al., 2010), mackerel (Catanese et al., 2010), oysters (Wang and Guo, 2008), salmon (Rasmussen et al., 2010), halibut, swordfish, sole, and Nile perch (Applewhite et al., 2012). In addition, multiplex PCR can be used for the identification of dried, raw, and cooked bonito (Lin et al., 2008a,b). The results from traditional PCR reactions (both single or multiplex) are visualized using agarose gels to verify that bands of expected size are generated during PCR.

Real-Time PCR

The development of real-time PCR has enabled scientists to visualize results of PCR reactions as they occur. Real-time PCR utilizes either specific fluorescent probes, such as TaqMan, or nonspecific fluorescent dyes, such as SYBR Green and Eva Green, that emit fluorescence when an amplicon is generated. These reagents can be applied directly to the PCR reactions, which allows for the amplification and real-time detection of target DNA fragments as the reaction progresses producing computer-generated results therefore eliminating the use of gel electrophoresis for end-point detection (Marmiroli et al., 2003). Real-time PCR is a robust, sensitive, rapid technique that is becoming a powerful tool in species identification in the seafood and aquaculture industries (Espineira and Vieites, 2015). Real-time PCR with TaqMan probes has been used to identify certain species of grouper (Trotta et al., 2005), flatfish, cod and squid (Herreo et al., 2010; Herrero et al., 2012a,b), Atlantic cod, haddock, and whiting (Taylor et al., 2002), tuna (Lopez and Pardo, 2005; Terio et al., 2010), salmon (Herrero et al., 2011), salmon and trout (Rasmussen et al., 2010), snappers and drums (Bayha et al., 2008) and for traceability of European eel (Espineira and Vietes, 2015). Real-time PCR with TaqMan probes or nonspecific dyes can also be used to quantitatively measure DNA in samples containing multiple species such as meat products (Anabe et al., 2007; Lago et al., 2011) and fish paste (Nagase et al., 2010).

Traditional PCR and real-time PCR offer several advantages such as low cost, simple procedures, and straightforward detection, including the ability to identify species in mixed samples. A disadvantage is that these methods require the design of diagnostic primers for each target species (Rasmussen-Hellberg and Morrissey, 2011). Real-time PCR in particular offers a reduction in contamination as the entire reaction is performed in a single tube. It can also be completed very quickly, with identification from a sample possible in under 1 hour on some systems. However, this method requires more factors to be considered in upfront assay design, including accounting for reaction efficiency.

RESTRICTION FRAGMENT LENGTH POLYMORPHISM

PCR-RFLP is another popular alternative to PCR sequencing and is based on analyzing polymorphisms in the lengths of particular restriction fragments of genetic code. As mentioned earlier, species-specific variations in the sizes of particular DNA fragments can often be analyzed by PCR amplification with visualization on an agarose gel. However, when the variations are <100 bp difference, it may be easier to view differences after the PCR amplicons are digested with endonucleases called restriction enzymes. This process results in species-specific restriction profiles (Liu and Cordes, 2004). In order to establish a protocol for species identification using PCR-RFLP, the target DNA fragment must be amplified by PCR and then sequenced to identify polymorphisms among the target species. Restriction enzymes are selected to recognize and cut specific sequences of DNA, resulting in a pattern of restriction fragments, which varies with species (Liu and Cordes, 2004). Once the sequence of the fragment has been established, the PCR amplicon of interest is simply digested with the preselected restriction enzymes and then its restriction pattern is compared with reference samples for species identification.

PCR-RFLP has been carried out with a variety of DNA regions for seafood species identification. The most commonly used DNA region, mt cyt b, has been used to identify numerous species such as scombroids (Chow et al., 2003; Horstkotte and Rehbein, 2003), flatfish (Sotelo et al., 2001), gadoids (Aranishi et al., 2005a,b; Pepe et al., 2005), salmonids (Russell et al., 2000), and eels (Aoyama et al., 2000). Other DNA fragments that have been analyzed by PCR-RFLP for species identification include: nuclear 5S rRNA, mt 16S rRNA,

COSIII, mt 12S rRNA, and *ATCO* to differentiate mackerel species (Aranishi, 2005), Atlantic salmon from rainbow trout (Carrera et al., 2000), scombroid species (Takeyama et al., 2001; Chow et al., 2003) and to identify various flatfish species (Comesana et al., 2003). As reported in the late 2000s and early 2010s, PCR-RFLP has been utilized to identify puffer fish (Hsieh et al., 2010), salmon (Rasmussen et al., 2010), fish-meal (Nagase et al., 2009), and various fish species in seafood samples (Nebola et al., 2010).

PCR-RFLP offers several advantages over other techniques. It is simpler than sequencing methods, rapid, reproducible, robust, and a laboratory technique that does not require expensive equipment (Aranishi, 2005). Numerous studies have shown that PCR-RFLP is suitable for analysis of closely related species, samples containing mixed species, and samples that have undergone various levels of processing, including heat sterilization (Applewhite et al., 2012; Lin and Hwang, 2007; Rea et al., 2009). However, the restriction digestion process adds time to the identification compared to traditional PCR, and represents an additional step where contamination or error can be introduced.

In PCR-RFLP, microfluidic, lab-on-a-chip technology utilizing capillary electrophoresis (CE) to analyze the DNA fragments can replace gel electrophoresis (Dooley et al., 2005a,b). After restriction digestion with a PCR-amplified DNA fragment, the fragments are loaded into a microchip, separated using CE, and then detected and quantified using laser-induced fluorescence (Dooley et al., 2005a,b). Lab-on-a-chip has been used in fish species identification studies to differentiate Atlantic salmon and rainbow trout and to identify a variety of whitefish species (Dooley et al., 2005a,b). Agilent Technologies (Santa Clara, CA) has a fish identification system based on PCR-RFLP and lab-on-a-chip technology with RFLP pattern matching software that can identify about 80 fish species (www.agilent.com/chem/fishID). The lab-on-a-chip technology also has the potential to quantify mixtures relative to the total DNA detected, enabling estimates of proportions where multiple species are present (Dooley et al., 2005b). While lab-on-a-chip CE may be more sensitive, faster, and more reliable than gel electrophoresis, it does not eliminate the inherent limitations of PCR-RFLP mentioned in the following section.

Although PCR-RFLP is a popular method in the field of seafood species identification, it does have some drawbacks. The possibility of intraspecies variations producing false results is a major disadvantage. This occurs when individuals from the same species show different restriction patterns due to degeneracy in the target DNA fragment (Akasaki et al., 2006). To identify potential intraspecies variations, numerous individuals from the same species must be analyzed at the target sites. Another complication with PCR-RFLP is that all species may not provide unique restriction patterns, thus an unknown species that has not yet been analyzed with PCR-RFLP could be falsely identified if its restriction profile matches that of a previously studied species (Sotelo et al., 2001). To overcome these limitations, it is recommended that species identification with PCR-RFLP be carried out with caution if there is not sufficient information available about sequence polymorphisms within and between the species groups (Mackie et al., 1999; Sotelo et al., 2001) and that at least two diagnostic restriction sites be used (Lenstra, 2003). In addition, banding patterns can be difficult to interpret if faint bands are present, and this can introduce some error in manual species identification based on the result of gel electrophoresis.

PCR SINGLE-STRANDED CONFORMATIONAL POLYMORPHISM

PCR single-stranded conformational polymorphism (PCR-SSCP) is another method for the detection of interspecies polymorphisms. RFLP has been reported to be simpler and more robust, however, SSCP is a highly sensitive technique that is less problematic than RFLP in regards to intraspecies variation (Mackie et al., 1999; Akasaki et al., 2006). As with RFLP, analysis begins with PCR amplification of a specific DNA fragment in all species being examined (Applewhite et al., 2012). The resultant amplicon is then denatured into a fragment of single-stranded DNA that has a secondary structure based on its sequence. Differences in sequence, even as small as a single nucleotide, can be detected by differences in electrophoretic mobility with polyacrylamide gel electrophoresis. SSCP patterns are visualized by silver staining and compared to the profiles of authentic species to identify an unknown sample (Mackie et al., 1999). SSCP is capable of analyzing small DNA fragments (~100bp) and also detecting species in processed samples (Rehbein et al., 2002; Rehbein, 2005) and mixed samples (Mackie et al., 1999; Rehbein et al., 1999b).

PCR-SSCP has been utilized to identify a variety of fish species, including salmonids, sardines, herring, eel, tuna, bonito, sturgeon (Rehbein et al., 1997, 1999a,b, 2002); clams (Fernandez et al., 2002); and species in the Pangasiidae family (Sriphairoj et al., 2010). Although successful in some species identification studies, SSCP analysis is more challenging than RFLP and has a number of limitations. PCR-SSCP is highly sensitive and requires a high level of reproducibility with no variation in conditions between analyses (Lockley and Bardsley, 2000). This can cause some issues in analysis at different facilities, for example. In addition, reference samples must always be run on the same gel as the unknown. This requires constant access to reference material from any species which the unknown sample will be compared to. Reference material can be difficult to obtain; however, promising results using Whole Genome Amplification as a means to generate additional reference material from a rare source suggests that it may be easier to make rare material last longer (Naaum et al., unpublished).

RANDOM-AMPLIFIED POLYMORPHIC DNA

In random-amplified polymorphic DNA (RAPD) analyses a short primer (~10nt in length) is designed and added to a PCR reaction with the target DNA. This arbitrary primer is designed without previous knowledge of the target DNA sequence and randomly amplifies segments of DNA in a PCR reaction (Williams et al., 1990). The PCR amplicons are then analyzed using gel electrophoresis. RAPD analysis on different species results in unique patterns of DNA fragments due to variations in the genetic code. If the resulting band patterns are species-specific, the DNA fingerprint for that species is established. Unknown samples can be analyzed using the same primer and their band patterns compared to the DNA fingerprints for known samples to verify the species.

RAPD has been used as an accurate, rapid tool for detecting commercial fraud (Ramella et al., 2005), and primers are commercially available (Lockley and Bardsley, 2000; Rego et al., 2002; Liu and Cordes, 2004). RAPD methods for species identification have been developed for catfish (Liu et al., 1998b), tilapia (Ahmed et al., 2004), mussels (Rego et al., 2002), Asian arowana (dragonfish: *Scleropages formosus*) (Yue et al., 2002), blackfin goosefish (*Lophius gastrophysus*) (Ramella et al., 2005), the giant grouper: *Epinephelus lanceolatus* (Chiu et al., 2012), and rainbow trout: *Oncorhynchus mykiss* (Afzali et al., 2013). However, most fish research with PCR-RAPD has been in population genetics rather than in species identification and commercial fraud (Ali et al., 2004).

RAPD has several advantages over RFPL and SSCP in that it does not require prior knowledge of the genome sequence for primer development, requires minimal DNA, and allows for both intra- and interspecies differentiation (Ramella et al., 2005). PCR-RAPD also has some disadvantages. One concern is reproducibility of the method, especially when the target DNA is degraded (Lockley and Bardsley, 2000; Rego et al., 2002). If the template DNA is of poor quality, some of the larger fragments common to specific fingerprints might not be present. Also, reaction conditions must be constant and rigorous, to ensure that the DNA fingerprints produced accurately mirror the corresponding species. Similar to RFLP, this can be problematic if samples analyzed at different laboratories produce different results for the same sample. Another complication can arise if different DNA regions from two species produce PCR fragments of the same length leading to misidentification (Liu and Cordes, 2004).

AMPLIFIED FRAGMENT LENGTH POLYMORPHISM

Amplified fragment length polymorphism (AFLP) is a DNA fingerprinting technique that has features of both RFLP and RAPD (Bensch and Akesson, 2005). AFLP analysis involves digestion of whole genomic DNA with two restriction enzymes. One restriction enzyme has a shorter sequence and cuts frequently while the other is a longer sequence and cuts less frequently. Adaptor molecules that recognize the restriction sequences are ligated to the DNA restriction fragments. PCR amplification is carried out with primers that anneal to the adaptor molecules. These primers contain an additional base at the 3′-end and only amplify a subset of the available DNA fragments (Bensch and Akesson, 2005).

The resulting amplicons are used as template DNA for a second PCR amplification that is more selective. This PCR step again reduces the number of available DNA fragments, resulting in a total of about 100 fragments. These fragments are separated by size using gel electrophoresis and detected by a fluorescent or radioactive label on the *Eco*RI adaptor-specific primer (Bensch and Akesson, 2005). The final result is a specific DNA fingerprint, where inter- and intra-species polymorphisms are indicated by the presence or absence of specific fragments.

AFLP is similar to RAPD in that it does not require prior knowledge of the DNA sequence, however, AFLP analysis shows greater levels of reproducibility and polymorphism (Bossier, 1999; Liu and Cordes, 2004). Although AFLP analysis results in numerous informative markers and complex banding patterns, information on individual DNA fragments is not as specific as with other techniques (Applewhite et al., 2012). Additionally, the development of AFLP markers is fairly labor-intensive and requires DNA of high quality and high molecular weight. This can limit its use for cooked, canned, or otherwise degraded food samples. While AFLP analysis has been extensively utilized for genetic research involving plants, fungi, and bacteria, it has not been used broadly in the field of animal research (Bensch and Akesson, 2005). AFLP markers have been developed for a few fish species (Han and Ely, 2002; Yu and Guo, 2003; Liu et al., 1998a; Li and Guo, 2004), but the majority of other studies utilizing AFLP are focused on the construction genetic linkage maps. AFLP is yet to be exploited in fish fraud research because it is relatively time-consuming compared to other methods and has not been adapted for large-scale applications (Zhang and Cai, 2006). Table 7.1 is a comparison of different DNA-based methods besides sequencing that are used in seafood species identification.

CONCLUSIONS

While all the methods discussed in this chapter offer varying advantages over sequencing techniques such as ease of the procedure, minimal costs associated with equipment and consumables, ability to identify multiple species in a sample, and the capacity to identify and differentiate species in processed foods, they all also have one major disadvantage in commercial species identification testing. These methods target specific species for which primers, probes, fingerprints, and so on must be developed or designed individually. These diagnostics can identify certain species, but unlike sequencing techniques, if the target species is not detected, these methods do not provide additional information as to the identity of the unknown species. There are over 1800 species of seafood listed on the FDA Seafood List as commercially sold in the United States. Therefore, a commercial testing laboratory would need a reference library of species-specific primers, probes, and/or fingerprints for all of these species to address mislabeled seafood and consumer fraud in commerce, and this would not include species that may be substituted that are not normally commercially

TABLE 7.1 Comparison of Major DNA-Based Methods Other Than Sequencing Used in Fish and Seafood Species Identification for Prevention of Commercial Fraud

DNA-Based Method	Acronym	Requires Prior DNA-Sequence Information?	Quantity of Loci Analyzed	Robustness to DNA Degradation	Potential for Interlaboratory Reproducibility	Cost	Potential for Database Construction	Potential for Intraspecies Variation Errors	Examples of Fish and Seafood Species Identified With Method
Species-specific primers and multiplex PCR	NA	Yes	Single	Medium to high	High	Medium	High	Medium	Flatfish, gadiformes, salmonids, scombroids, percoids, sturgeon, eels, sharks, mollusks
Restriction fragment length polymorphism	RFLP	Yes	Single	Medium to high	High	Medium	Medium to high	Medium	Flatfish, gadiformes, salmonids, scombroids, percoids, sturgeon, eels, mollusks
Single-stranded conformational polymorphism	SSCP	Yes	Single	Medium to high	Medium	Medium	Medium to high	Low to medium	Salmonids, scombroids, sturgeon, eels
Random-amplified polymorphic DNA	RAPD	No	Multiple	Low to medium	Low to medium	Medium	Medium to high	Low to medium	Percoids, goosefish, mollusks
Amplified fragment length polymorphism	AFLP	No	Multiple	Low to medium	Medium to high	Medium to high	Medium to high	Low to medium	Salmonids, scombroids

The terms "low", "medium", and "high" in the table are descriptions designated by Rasmussen and Morrissey of their review of the various methodologies. Modified and printed with permission from Rasmussen, R., Morrissey, M.T., 2008. DNA-based methods for the identification of commercial fish and seafood species. Comprehensive Reviews in Food Science and Food Safety 7 (3), 280–295.

traded, and therefore not represented on this list. However, the species-specific techniques discussed in this chapter do represent significant benefit for the identification of specific target species in terms of cost and time required for identification, and allow analysis of mixed or processed samples.

EMERGING TECHNOLOGIES

DNA chips (also known as DNA microarrays or DNA macroarrays) are tools made by spotting hundreds to tens of thousands of genetic samples known as probes onto a solid support matrix. The spots can be DNA, cDNA, or oligonucleotides. Typically, the probes used to construct a DNA chip are specific for a variety of sequences (DNA or mRNA) within a single organism; however, probes can also be designed to detect sequences across multiple organisms.

Labeled target sequences (typically using fluorescence) bind to a probe sequence to generate a signal on the chip. Total strength of the signal, from a spot (feature), depends upon the amount of target sample binding to the probes present on that spot.

These assays are a valuable tool in several fields because they have the potential to simultaneously identify up to hundreds or thousands of species (Teletchea et al., 2005). For aquatic species, initial DNA chips that were developed were used primarily for ecotoxicology research (Larkin et al., 2002, 2003; Brown et al., 2004; Hoyt et al., 2003) and were predominantly constructed by shotgun-sequence fragments of genes from cDNA libraries.

Since the initial construction of these DNA chips, there has been an exponential increase in the availability of sequence information for a variety of species, which can be used to construct DNA chips for other species. Genome information that can be used to construct high-density microarrays is available at the National Center for Biotechnology Information (http://www.ncbi.nlm. nih.gov/) for a number of aquatic organisms including Atlantic salmon (*Salmo salar*), Zebrafish (*Danio rerio*), fathead minnow (*Pimephales promelas*), killifish (*Fundulus heteroclitus*), and other fish. Most of the previously mentioned species are used as models in research rather than being commercially important in the seafood industry. The one exception is Atlantic salmon, whose genome is currently being sequenced and assembled by the International Cooperation to Sequence the Atlantic Salmon Genome (ICSASG) group (http://www.icisb.org/ atlantic-salmon-genome-sequence/). The genome of Atlantic Salmon has been currently assembled to the "chromosome level," which is the second highest sequence assembly level below a "complete genome" where all chromosomes are gapless and have no runs of 10 or more ambiguous bases.

DNA chips are starting to be developed and utilized in the seafood industry. Kochzius et al. (2010) developed and tested an oligonucleotide microarray containing probes for the mitochondrial genes 16S rRNA (16S), cytochrome *b* (cyt *b*), and cytochrome oxidase subunit I (COI). The microarray was used to identify 30 European marine fish species.

An oligonucleotide-based DNA chip was also developed to differentiate six jellyfish species using the COI gene (Lee et al., 2011). A DNA chip was developed to differentiate six animal species commonly consumed in Europe (Peter et al., 2004). Universal primers were used to amplify a 377-bp fragment of the mt cyt *b* gene. The resulting fragments were then identified in a microarray utilizing species-specific oligonucleotide probes. This DNA chip was able to detect species that were present at 0.1% and could identify up to four species simultaneously in mixed commercial food samples.

A commercial DNA chip-based product called the FoodExpert-ID was launched in France in 2004 by BioMerieux (http://www.biomerieux.com). According to the Website, this product contained the first high-density DNA chip for use with species identification in food and animal feeds, and is able to detect 33 different species of vertebrates, including 15 species of fish. While used in Europe, the company has not launched the product in the US and has no plans to do so at this time. Another DNA microarray was developed in 2008 to differentiate 11 commercially important fish species based on a 600-bp fragment of the 16S rDNA gene (Kochzius et al., 2008). A "Fish Chip" for the identification of approximately 50 species found in European Seas has been developed for authentication and research purposes in the fisheries industry. Array-based methods have not been investigated extensively for species identification in seafoods. This may be in part due to high development costs and overall expense of running arrays.

Another emerging molecular tool being utilized to identify multiple species at a time in the seafood field includes a study by Gleason and coworkers who developed oligonucleotide probes for 23 marine fish species that produce pelagic eggs commonly found in California waters. These probes were coupled to fluorescent microspheres to create a suspension bead array. Biotin-labeled primers were used to amplify the 16S ribosomal rRNA gene and the mitochondrial COI gene from individual fish eggs. The PCR resultant PCR amplicons were hybridized to the bead array, and following the addition of a reporter fluorophore, samples were analyzed by flow cytometry (Gleason and Burton, 2012).

Another technology based on the design of species-specific primers, isothermal amplification, and lateral flow detection has been developed by TwistDx (http://www.twistdx.co.uk/). This technology employs isothermal recombinase polymerase amplification (RPA). The RPA procedure involves recombinase enzymes, which can pair oligonucleotide primers with homologous DNA sequences. A strand displacing DNA polymerase can then extend from the primer-bound complex to synthesize a new complementary DNA strand. Similar to PCR, the use of two opposing primers allows exponential amplification of the target sequence. RPA probes are designed to hybridize to a target sequence within the amplicon. Detection of the resultant amplicon can be via real-time fluorescence (Fpg or Exo probes) or by immunochromatographic strip ICS detection (Nfo probes). This procedure does not require a PCR machine and operates most efficiently at 37–42°C. This technology has been developed for the diagnosis of different pathogens (Abd El Wahed et al., 2013a,b; Ahmed et al., 2015; Amer et al., 2013; Piepenburg et al., 2006). The company also offers a Red Snapper kit to identify *Lutjanus campechanus*.

REFERENCES

Abd El Wahed, A., El-Deeb, A., El-Tholoth, M., Abd El Kader, H., Ahmed, A., Hassan, S., Hoffmann, B., Haas, B., Shalaby, M.A., Hufert, F.T., Weidmann, M., 2013a. A portable reverse transcription recombinase polymerase amplification assay for rapid detection of foot-and-mouth disease virus. PLoS One 8, e71642.

Abd El Wahed, A., Patel, P., Heidenreich, D., Hufert, F.T., Weidmann, M., 2013b. Reverse transcription recombinase polymerase amplification assay for the detection of middle East respiratory syndrome coronavirus. PLoS Currents 5, 1–11.

Afzali, B., Mianji, G., Gholizadeh, M., 2013. Genetic variability of rainbow trout (*Oncorhynchus mykiss*) cultured in Iran using molecular RAPD markers. Iranian Journal of Fisheries Sciences 12 (3), 511–521.

Ahmed, M.M.M., Ali, B.A., El-Zaeem, S., 2004. Application of RAPD markers in fish: part I – Some genera (*Tilapia, Sarotherodon* and *Oreochromis*) and species (*Oreochromis aureus* and *Oreochromis niloticus*) of tilapia. International Journal of Biotechnology 6, 86–93.

Ahmed, S.A., van de Sande, W.W.J., Desnos-Olliver, M., Fahal, A.H., Mhmoud, N.A., de Hoog, G.S., 2015. Application of isothermal amplification techniques for identification of *Madurella mycetomatis*, the prevalent agent of human mycetoma. Journal of Clinical Microbiology 53 (1), 3280–3285.

Akasaki, T., Yanagimoto, T., Yamakami, K., Tomonaga, H., Sato, S., 2006. Species identification and PCR-RFLP analysis of cytochrome *b* gene in cod fish (order *Gadiformes*) products. Journal of Food Science 71 (3), C190–C195.

Ali, B.A., Huang, T.H., Qin, D.N., Wang, X.M., 2004. A review of random amplified polymorphic DNA (RAPD) markers in fish research. Reviews in Fish Biology and Fisheries 14, 443–453.

Amer, H.M., Abd El Wahed, A., Shalaby, M.A., Almajhdi, F.N., Hufert, F.T., Weidmann, M., 2013. A new approach for diagnosis of bovine coronavirus using a reverse transcription recombinase polymerase amplification assay. Journal of Virological Methods 193, 337–340.

Anabe, S., Hase, M., Yano, T., Sato, M., Fujimura, T., Akiyama, H., 2007. A real-time quantitative PCR detection method for pork, chicken, beef, mutton, and horseflesh in foods. Bioscience, Biotechnology, and Biochemistry 71, 3131–3135.

Aoyama, J., Watanabe, S., Nishida, M., Tsukamoto, K., 2000. Discrimination of catadromous eels of genus Anguilla using polymerase chain reaction-restriction fragment length polymorphism analysis of the mitochondrial 16S ribosomal RNA domain. Transactions of the American Fisheries Society 129 (3), 873–878.

Applewhite, L., Rassmussen, R., Morrissey, M., 2012. Species identification of seafood. In: The Seafood Industry: Species, Products, Processing, and Safety, second ed. Wiley-Blackwell, pp. 193–229.

Apte, A., Daniel, S., 2003. PCR primer design. In: Dieffenbach, C.W., Dveksler, G.S. (Eds.), PCR Primer: A Laboratory Manual, second ed. Cold Spring Harbor, New York, pp. 61–74.

Aranishi, F., Okimoto, T., Izumi, S., 2005a. Identification of gadoid species (Pisces, Gadidae) by PCR-RFLP analysis. Journal of Applied Genetics 46 (1), 69–73.

Aranishi, F., Okimoto, T., Ohkubo, M., Izumi, S., 2005b. Molecular identification of commercial spicy pollack roe products by PCR-RFLP analysis. Journal of Food Science 70 (4), C235–C238.

Aranishi, F., 2005. Rapid PCR-RFLP method for discrimination of imported and domestic mackerel. Marine Biotechnology 7 (6), 571–575.

Bauer, T., Weller, P., Hammes, W., Hertel, C., 2003. The effect of processing parameters on DNA degradation in Food. European Food Research and Technology 217, 338–343.

Bayha, K.M., Graham, W.M., Hernandez, F.J., 2008. Multiplex assay to identify eggs of three fish species from the northern Gulf of Mexico, using locked nucleic acid Taqman real-time PCR probes. Aquatic Biology 4, 65–73.

Bensch, S., Akesson, M., 2005. Ten years of AFLP in ecology and evolution: why so few animals? Molecular Ecology 14 (10), 2899–2914.

Bossier, P., 1999. Authentication of seafood products by DNA patterns. Journal of Food Science 64 (2), 189–193.

Brown, M., Robinson, C., Davies, I.M., Moffat, C.F., Redshaw, J., Craft, J.A., 2004. Temporal changes in gene expression in the liver of male plaice (*Pleuronectes platessa*) in response to exposure to ethynyl oestradiol analysed by macroarray and real-time PCR. Mutation Research 552, 35–49.

Carrera, E., Garcia, T., Cespedes, A., Gonzalez, I., Fernandez, A., Asensio, L.M., Hernandez, P.E., Martin, R., 2000. Identification of smoked Atlantic salmon (*Salmo salar*) and rainbow trout (*Oncorhynchus mykiss*) using PCR-restriction fragment length polymorphism of the *p53* gene. Journal of AOAC International 83 (2), 341–346.

Catanese, G., Manchado, M., Fernandez-Trujillo, A., Infante, C., 2010. A multiplex-PCR assay for the authentication of mackerels of the genus *Scomber* in processed fish products. Food Chemistry 122, 319–326.

Chiu, T., Su, Y., Pa, U., Chang, H., 2012. Molecular markers for detection and diagnosis of the giant grouper (*Epinephelus lanceolatus*). Food Control 24, 29–37.

Chow, S., Nohara, K., Tanabe, T., Itoh, T., Tsuji, S., Nishikawa, Y., Uyeyanagi, S., Uchikawa, K., 2003. Genetic and morphological identification of larval and small juvenile tunas (Pisces: Scombridae) caught by a mid-water trawl in the western Pacific. Bulletin of Fisheries Research Agency 8, 1–14.

Comesana, A.S., Abella, P., Sanjuan, A., 2003. Molecular identification of five commercial flatfish species by PCR-RFLP analysis of a 12S rRNA gene fragment. Journal of the Science of Food and Agriculture 83, 752–759.

Dooley, J.J., Sage, H.D., Brown, H.M., Garrett, S.D., 2005a. Improved fish species identification by use of lab-on-a-chip technology. Food Control 16, 601–607.

Dooley, J.J., Sage, H.D., Clarke, M.A., Brown, H.M., Garrett, S.D., 2005b. Fish species identification using PCR-RFLP analysis and lab-on-a-chip capillary electrophoresis: application to detect white fish species in food products and an interlaboratory study. Journal of Agricultural and Food Chemistry 53 (9), 3348–3357.

Espineira, M., Vieites, J., 2015. Genetic system for an integral traceability of European eel (*Anguilla anguilla*) in aquaculture and seafood products: authentication by fast real-time PCR. European Food Research and Technology. http://dx.doi.org/10.1007/s00217-015-2514-y.

Fernandez, A., Garcia, T., Gonzalez, I., Asensio, L., Rodriguez, M.A., Hernandez, P.E., Martin, R., 2002. Polymerase chain reaction-restriction fragment length polymorphism analysis of a 16S rRNA gene fragment for authentication of four clam species. Journal of Food Protection 65 (4), 692–695.

Gleason, L.U., Burton, R.S., 2012. High-throughput molecular identification of fish eggs using multiplex suspension bead arrays. Molecular Ecology Resources 12, 57–66.

Han, K., Ely, B., 2002. Use of AFLP analyses to assess genetic variation in *Morone* and *Thunnus* species. Marine Biotechnology 4 (2), 141–145.

Herrero, B., Madrinán, M., Vieites, J.M., Espiñeira, M., 2010. Authentication of Atlantic cod (*Gadus morhua*) using real time PCR. Journal of Agricultural and Food Chemistry 58 (8), 4794–4799.

Herrero, B., Vieites, J.M., Espiñeira, M., 2011. Authentication of Atlantic salmon (*Salmo salar*) using real-time PCR. Food Chemistry 127 (3), 1268–1272.

Herrero, B., Lago, F.C., Vieites, J.M., Espiñeira, M., 2012a. Rapid method for controlling the correct labeling of products containing European squid (*Loligo vulgaris*) by fast real-time PCR. European Food Research and Technology 234 (1), 77–85.

Herrero, B., Lago, F.C., Vieites, J.M., Espiñeira, M., 2012b. Real-time PCR method applied to seafood products for authentication of European sole (*Solea solea*) and differentiation of common substitute species. Food Additives & Contaminants: Part A 29 (1), 12–18.

Horstkotte, B., Rehbein, H., 2003. Fish species identification by means of restriction fragment length polymorphism and high-performance liquid chromatography. Journal of Food Science 68 (9), 2658–2666.

Hoyt, P.R., Doktycz, M.J., Beattie, K.L., Greeley, M.S.J., 2003. DNA microarrays detect 4-nonylphenol-induced alterations in gene expression during zebrafish early development. Ecotoxicology 12, 469–474.

Hsieh, C.H., Chang, W.T., Chang, H.C., Hsieh, H.S., Chung, Y.L., Hwang, D., 2010. Puffer fish-based commercial fraud identification in a segment of cytochrome b region by PCR-RFLP analysis. Food Chemistry 121 (4), 1305–1311.

Kochzius, M., Nölte, M., Weber, H., Silkenbeumer, N., Hjörleifsdottir, S., Hreggvidsson, G.O., Marteinsson, V., Kappel, K., Planes, S., Tinti, F., Magoulas, A., Garcia Vazquez, E., Turan, C., Hervet, C., Campo Falgueras, D., Antoniou, A., Landi, M., Blohm, D., 2008. DNA microarrays for identifying fishes. Marine Biotechnology 10, 207–217.

Kochzius, M., et al., 2010. Identifying fishes through DNA barcodes and microarrays. PLoS One 5 (9), e12620.

Lago, F.C., Herrero, B., Madriñán, M., Vieites, J.M., Espiñeira, M., 2011. Authentication of species in meat products by genetic techniques. European Food Research and Technology 232 (3), 509–551.

Larkin, P., Sabo-Attwood, T., Kelso, J., Denslow, N.D., 2002. Gene expression analysis of largemouth bass exposed to estradiol, nonylphenol, and p,p'-DDE. Comparative Biochemistry and Physiology Part B 133, 543–557.

Larkin, P., Folmar, L.C., Hemmer, M.J., Poston, A.J., Denslow, N.D., 2003. Expression profiling of estrogenic compounds using a sheepshead minnow cDNA macroarray. Environmental Health Perspectives 111, 29–36.

Lee, G., Park, S.Y., Hwang, J., Lee, Y., Hwang, S.Y., Lee, S., Lee, T., 2011. Development of DNA chip for jellyfish verification from South Korea. BioChip 5 (4), 375–382.

Lenstra, J.A., 2003. DNA methods for identifying plant and animal species in food. In: Lees, M. (Ed.), Food Authenticity and Traceability. Woodhead Publishing Limited, Cambridge, England, pp. 34–53.

Li, L., Guo, X., 2004. AFLP-based genetic linkage maps of the pacific oyster *Crassostrea gigas* Thunberg. Marine Biotechnology 6 (1), 26–36.

Lin, W.F., Hwang, D.F., 2007. Application of PCR-RFLP analysis on species identification of canned tuna. Food Control 18, 1050–1057.

Lin, W.F., Hwang, D.F., 2008a. Application of species-specific PCR for the identification of dried bonito product (Katsuobushi). Food Chemistry 106, 390–396.

Lin, W.F., Hwang, D.F., 2008b. A multiplex PCR assay for species identification of raw and cooked bonito. Food Control 19 (9), 879–885.

Liu, Z.J., Cordes, J.F., 2004. DNA marker technologies and their applications in aquaculture genetics. Aquaculture 238, 1–37.

Liu, Z., Nichols, A., Li, P., Dunham, R.A., 1998a. Inheritance and usefulness of AFLP markers in channel catfish (*Ictalurus punctatus*), blue catfish (*I. furcatus*), and their F1, F2, and backcross hybrids. Molecular Genetics and Genomics 258 (3), 260–268.

Liu, Z.J., Li, P., Argue, B., Dunham, R., 1998b. Inheritance of RAPD markers in channel catfish (*Ictalurus punctatus*), blue catfish (*I. furcatus*), and their F1, F2 and backcross hybrids. Animal Genetics 29, 58–62.

Lockley, A.K., Bardsley, R.G., 2000. DNA-based methods for food authentication. Trends in Food and Science Technology 11, 67–77.

Lopez, I., Pardo, M.A., 2005. Application of relative quantification TaqMan real-time polymerase chain reaction technology for the identification and quantification of *Thunnus alalunga* and *Thunnus albacares*. Journal of Agricultural and Food Chemistry 53 (11), 4554–4560.

Mackie, I.M., Pryde, S.E., Gonzales-Sotelo, C., Medina, I., Perez-Martin, R.I., Quinteiro, J., Rey-Mendez, M., Rehbein, H., 1999. Challenges in the identification of species of canned fish. Trends in Food and Science Technology 10, 9–14.

Mafra, I., Ferreira, I.M., Oliveira, M.B., October 30, 2007. Food authentication by PCR-based methods. European Food Research and Technology.

Marmiroli, N., Peano, C., Maestri, E., 2003. Advanced PCR techniques in identifying food components. In: Lees, M. (Ed.), Food Authenticity and Traceability. Woodhead Publishing Limited, Cambridge, England, pp. 3–33.

Mendonc, A.F.F., Hashimoto, D.T., De-Franco, B., Porto-Foresti, F., Gadig, O.B.F., Oliveira, C., Foresti, F., 2010. Genetic identification of lamniform and carcharhiniform sharks using multiplex-PCR. Conservation Genetics Resources 2 (1), 31–35.

Nagase, M., Maeta, K., Aim, T., Suginaka, K., Morinaga, T., 2009. Authentication of flying-fish-meal content of processed food using PCR-RFLP. Fisheries Science 75 (3), 811–816.

Nagase, M., Ye, R., Hidaka, F., Maeta, K., Aimi, T., Yamaguchi, T., Suginaka, K., Morinaga, T., 2010. Quantification of relative flying fish paste content in the processed seafood ago-noyaki using real-time PCR. Fisheries Science 76, 885–892.

Nebola, M., Borilova, G., Kasalova, J., 2010. PCR-RFLP analysis of DNA for the differentiation of fish species in seafood samples. Bulletin of the Veterinary Institute in Pulawy 54, 49–53.

Pardo, M.A., Perez-Villareal, B., 2004. Identification of commercial canned tuna species by restriction site analysis of mitochondrial DNA products obtained by nested primer PCR. Food Chemistry 86, 143–150.

Pepe, T., Trotta, M., di Marco, I., Cennamo, P., Anastasio, A., Cortesi, M.L., 2005. Mitochondrial cytochrome *b* DNA sequence variations: an approach to fish species identification in processed fish products. Journal of Food Protection 68 (2), 421–425.

Peter, C., Brunen-Nieweler, C., Cammann, K., Borchers, T., 2004. Differentiation of animal species in food by oligonucleotide microarray hybridization. European Food Research and Technology 219, 286–293.

Piepenburg, O., Williams, C.H., Stemple, D.L., Armes, N.A., 2006. DNA detection using recombination proteins. PLoS Biology 4, e204.

Ramella, M.S., Kroth, M.A., Tagliari, C., Arisi, A.C.M., 2005. Optimization of random amplified polymorphic DNA protocol for molecular identification of *Lophius gastrophysus*. Ciência e Tecnologia de Alimentos 25 (4), 733–735.

Rasmussen, R., Morrissey, M.T., 2008. DNA-based methods for the identification of commercial fish and seafood species. Comprehensive Reviews in Food Science and Food Safety 7 (3), 280–295.

Rasmussen, R.S., Morrissey, M.T., Walsh, J., 2010. Application of a PCR-RFLP method to identify salmon species in U.S. commercial products. Journal of Aquatic Food Product Technology 19, 3–15.

Rasmussen-Hellberg, R.S., Morrissey, M.T., 2011. Advances in DNA-based techniques for the detection of seafood species substitution on the commercial market. JALA 308–321.

Rea, S., Storani, G., Mascaro, N., Stocchi, R., Loschi, A.R., 2009. Species identification in anchovy pastes from the market by PCR RFLP technique. Food Control 20 (5), 515–520.

Rego, I., Martinez, A., Gonzalez-Tizon, A., Vieites, J., Leira, F., Mendez, J., 2002. PCR technique for the identification of mussel species. Journal of Agricultural and Food Chemistry 50 (7), 1780–1784.

Rehbein, H., Kress, G., Schmidt, T., 1997. Application of PCR-SSCP to species identification of fishery products. Journal of the Science of Food and Agriculture 74, 35–41.

Rehbein, H., Gonzales-Sotelo, C., Perez-Martin, R.I., Quinteiro, J., Rey-Mendez, M., Pryde, S., Mackie, I.M., Santos, A.T., 1999a. Differentiation of sturgeon caviar by single strand conformational polymorphism (PCR-SSCP) analysis. Archiv für Lebensmittelhygiene 50, 13–17.

Rehbein, H., Mackie, I.M., Pryde, S., Gonzales-Sotelo, C., Medina, I., Perez-Martin, R.I., Quinteiro, J., Rey-Mendez, M., 1999b. Fish species identification in canned tuna by PCR-SSCP: validation by a collaborative study and investigation of intra-species variability of the DNA-patterns. Food Chemistry 64 (2), 263–268.

Rehbein, H., Sotelo, C.G., Perez-Martin, R.I., Chapela-Garrido, M.J., Hold, G.L., Russell, V.J., Pryde, S., Santos, A.T., Rosa, C., Quinteiro, J., Rey-Mendez, M., 2002. Differentiation of raw or processed eel by PCR-based techniques: restriction fragment length polymorphism analysis (RFLP) and single strand conformation polymorphism analysis (SSCP). European Food Research and Technology 214 (2), 171–177.

Rehbein, H., 2005. Identification of the fish species of raw or cold-smoked salmon and salmon caviar by single-strand conformation polymorphism (SSCPP) analysis. European Food Research and Technology 220, 625–632.

Russell, V.J., Hold, G.L., Pryde, S.E., Rehbein, H., Quinteiro, J., Rey-Mendez, M., Sotelo, C.G., Perez-Martin, R.I., Santos, A.T., Rosa, C., 2000. Use of restriction fragment length polymorphism to distinguish between salmon species. Journal of Agricultural and Food Chemistry 48 (6), 2184–2188.

Sotelo, C.G., Calo-Mata, P., Chapela, M.J., Perez-Martin, R.I., Rehbein, H., Hold, G.L., Russell, V.J., Pryde, S., Quinteiro, J., Izquierdo, M., Rey-Mendez, M., Rosa, C., Santos, A.T., 2001. Identification of flatfish (*Pleuronectiforme*) species using DNA-based techniques. Journal of Agricultural and Food Chemistry 49 (10), 4562–4569.

Sriphairoj, K., Klinbunga, S., Kamonrat, W., Na-Nakorn, U., 2010. Species identification of four economically important Pangasiid catfishes and closely related species using SSCP markers. Aquaculture 308, S47–S50.

Takeyama, H., Chow, S., Tsuduki, H., Matsunaga, T., 2001. Mitochondrial DNA sequence variation within and between *Thunnus* tuna species and its application to species identification. Journal of Fish Biology 58 (6), 1646–1657.

Taylor, M.I., Fox, C., Rico, I., Rico, C., 2002. Species-specific TaqMan probes for simultaneous identification of (*Gadus morhua* L.), haddock (*Melanogrammus aeglefinus* L.) and whiting (*Merlangius merlangus* L.). Molecular Ecology Notes 2, 599–601.

Teletchea, F., Maudet, C., Hanni, C., 2005. Food and forensic molecular identification: update and challenges. Trends in Biotechnology 23 (7), 359–366.

Terio, V., Di Pinto, P., Decaro, N., Parisi, A., Desario, C., Martella, V., Buonavoglia, C., Tantillo, M., 2010. Identification of tuna species in commercial cans by minor groove binder probe real-time polymerase chain reaction analysis of mitochondrial DNA sequences. Molecular and Cellular Probes 24, 352–356.

Trotta, M., Schonhuth, S., Pepe, T., Cortesi, M.L., Puyet, A., Bautista, J.M., 2005. Multiplex PCR method for use in real-time PCR for identification of fish fillets from grouper (*Epinephelus* and *Mycteroperca* species) and common substitute species. Journal of Agricultural and Food Chemistry 53 (6), 2039–2045.

Wang, H., Guo, X., 2008. Identification of Crassostrea ariakensis and related oysters by multiplex species-specific PCR. Journal of Shellfish Research 27 (3), 481–487.

Williams, J.G., Kubelik, A.R., Livak, K.J., Rafalski, J.A., Tingey, S.V., 1990. DNA polymorphisms amplified by arbitrary primers are useful as genetic markers. Nucleic Acids Research 18 (22), 6531–6535.

Yu, Z., Guo, X., 2003. Genetic linkage map of the eastern oyster *Crassostrea virginica* Gmelin. Biological Bulletin 204 (3), 327–338.

Yue, G., Li, Y., Chen, F., Cho, S., Lim, L.C., Orban, L., 2002. Comparison of three DNA marker systems for assessing genetic diversity in Asian arowana (*Scleropages formosus*). Electrophoresis 23 (7–8), 1025–1032.

Zhang, J., Cai, Z., 2006. Differentiation of the rainbow trout (*Oncorhynchus mykiss*) from Atlantic salmon (*Salmo salmar*) by the AFLP-derived SCAR. European Food Research and Technology 223, 413–417.

Chapter 8

Population or Point-of-Origin Identification

Einar Eg Nielsen
Technical University of Denmark, Silkeborg, Denmark

WHY POPULATION OR POINT-OF-ORIGIN FOR SEAFOOD AUTHENTICITY AND TRACEABILITY

Population or point-of-origin identification represents the intermediate step in a continuum of DNA-based seafood authenticity and traceability applications, ranging from documenting the species to identifying the specific individual present in a product. While DNA methods have gained wide acceptance and application for species identification, population or point-of-origin assignment has received less attention and thus found fewer practical applications. The main reason for this is that these analyses require more background genetic data, more advanced statistical analysis, and a broader insight into the evolution and biology of the focal organisms to interpret the analytical output. However, the obstacles for wider implementation can now be overcome more easily, and point-of-origin identification is beginning to be implemented.

There are many good reasons for engaging in DNA-based population/point-of-origin identification for seafood. First, most seafood resource management is based on a system of spatially defined species-specific "stocks," outlined by international organization, such as FAO, ICES, and NAFO, and supported by fisheries legislation. In order to assure sustainability, the resource is assessed within each area and fishing quotas are given based on stock status. Illegal, unreported, and unregulated (IUU) fishing poses a significant threat to good management through local or regional overfishing and depletion of seafood resources. Thus there is a need for tools that can identify and document that the raw material entering the seafood production chain originates from sustainable fisheries. Secondly, many seafood products can originate from both wild capture fisheries and aquaculture with associated product differences with respect to quality, environmental impact, and animal welfare. Thirdly, seafood from different regions may vary with respect to classical measures of quality as, for example, taste, fat content, and color (Børresen, 1992), but also health aspects such as content of essential fatty acids, heavy

Seafood Authenticity and Traceability. http://dx.doi.org/10.1016/B978-0-12-801592-6.00008-5
149

metals, and frequency of disease-carrying agents. Collectively, quality measures for seafood products from a specific region, termed "terroir," may be labeled and branded to obtain a higher market price. Finally, there may be personal reasons, including political, why consumers would care about the origin of seafood products. For example, certain consumers would preferentially buy local products to support regional food production or to minimize CO_2 footprints of transporting seafood. To enable consumer choice with respect to these factors, seafood must be accurately identified and labeled.

Based on the considerations discussed earlier, the need for information on the origin of seafood is widely recognized and reflected in international laws. For example, in the European Union catch certificates that state the origin of all traded fish and fish products are required through the European Commission Control and IUU Regulations (EC, 2008, 2009). In addition to legal requirements, voluntary ecolabels, such as the MSC (Marine Stewardship Council, www.msc.org), for seafood products from capture fisheries have emerged, where one of the major pillars regarding certification of fisheries is sustainable fish stocks. A parallel system, ASC (Aquaculture Stewardship Council, www.asc.org) has emerged for aquaculture products, which among other aspects, takes biodiversity, sustainability and both fish and consumer health issues into account. Traditionally, compliance with rules and regulations, as well as voluntary ecolabeling, has to a very large extent relied on a paper trail documenting the origin of the product throughout the seafood supply chain. However, evidence from incidences in other parts of the food sector, for example, through the European BSE scandal has raised the awareness that there is a definite need for independent methods that are not easily fabricated that can validate the paper based traceability scheme.

A number of different methods have been suggested and applied independently or in concert (Higgins et al., 2010) for tracing the origin of seafood. These include morphometrics, meristics, micro/macroparasites, chemical composition, and analysis of fatty acids (Cadrin et al., 2014). Despite some success for inferring the origin of seafood, these methods are often hampered by limited availability of tissue of sufficient quality. This is particularly problematic for the analysis of processed seafood where most of these methods cannot be applied. Furthermore, calibration and standardization among laboratories necessary for forensic purposes (Ogden, 2008; Ogden and Linacre, 2015) is inherently difficult. For example, the establishment of long-term baseline datasets, to which any new specimen or products can be compared, is notoriously difficult. Finally, the statistical power for origin assignment associated with these methods individually is generally low. Thus concerted application of many methods requiring diverse expertise and equipment is beyond the capacity of most seafood laboratories and rarely applicable in a close to real-time framework required for practical seafood identification on a commercial scale, for example, for production line testing.

In contrast to biological- and chemical-based methods for seafood traceability, genetic methods offer many advantages. First, DNA is found in almost all cells in all organisms and can be retrieved from degraded or processed material, that is, DNA analysis can in principle be conducted at all stages in the production chain from sampling fresh fish onboard vessels to a filet on a dinner plate in a restaurant. Likewise, DNA-based origin assignment relies on a well-established theoretical framework from population and evolutionary genetics, allowing comparisons of new samples to already established genetic databases and at the same time providing a robust statistical framework for evaluating the result, which is essential for forensic purposes. In this context, the calibration of DNA results across laboratories is much simpler than for other methods, allowing global genetic information for various species to be compiled and used across laboratories without the need first to establish a new genetic baseline database. This, in turn, allows swift processing of new samples in a more real-time framework, as only the specific new "case" samples have to be processed before inferences on their geographical origin can be provided. Naturally, there are also limitations to the use of DNA analysis for origin assignment. In contrast to species designation, where mtDNA sequence information is available for all commercially valuable seafood species, databases of genetic information on populations across the species distribution are either incomplete or completely missing for many species. Furthermore, management areas and genetic populations are often not aligned (see Reiss et al., 2009). Management areas may include several genetic populations, or complicating matters further, the same population may be found in different management areas. To a large extent, this reflects that management areas have been defined on a political rather than a biological background. Still, the problems of mismatch between population and management area can be remedied in many cases. This is elaborated further in the following section.

Overall there continues to be a need for population/origin assignment in the seafood industry, and DNA-based methods represent the most promising means for practical use today. There are many very good reasons for using DNA-based methods for population/origin assignment. This chapter provides (1) an introduction to population genetics of marine organisms, (2) a review of the basic principles of population/origin identification, (3) a description of the various methods applied, (4) case studies of population and origin assignment, (5) a summary of caveats and potential pitfalls, and finally (6) a review of the ongoing and expected future developments within the field.

POPULATION GENETICS OF MARINE ORGANISMS

What Is a Genetic Population?

Genetic point-of-origin identification is based on assigning fish back to their genetic or evolutionary population. Under an evolutionary paradigm, a population can be defined as "A group of individuals of the same species living in close enough proximity that any member of the group can potentially mate with any other member" (Waples and Gaggiotti, 2006). That definition is distinct from an ecological paradigm where a population is defined as individuals of the same species that cooccur in an area and potentially interact. Thus the distinction is related to reproduction and a genetic population should therefore be reproductively isolated to some degree from any other genetic population of the species. The term "to some degree" is, however, quite vague and hardly operational. A more quantitative and practical definition of when groups of individuals are different enough to be considered populations is based on the exchange of effective migrants between populations per generation ($N_e m$, see later). If this number is above 25, populations become very difficult to distinguish using standard genetic tools (Waples and Gaggiotti, 2006). So, when populations are practically genetically indistinguishable they are, in this framework, defined as a single population. This distinction between theoretical delimitations of populations to a more application-based definition is important in applying DNA-based identification of populations to seafood.

Evolutionary Forces and Genetic Population Structure

Following evolutionary theory, individuals within reproductively isolated populations are subject to the same evolutionary forces that determine their genetic composition. These are: mutation, migration, random genetic drift, and selection. In popular terms, mutation is the long-term process that generates the genetic raw material in the form of new genetic variants, "alleles" at any gene locus (position of the DNA sequence in question in the genome). Migration of individuals carries those alleles among populations, and if migrants are successfully interbreeding, spread them through the process of "gene flow". Random genetic drift is the sampling error associated with breeding; that is, if "effectively" few individuals (where N_e is defined as the effective population size) participate in mating, then allele frequencies in the population will change fast and ultimately lead to the loss of allelic variants. Finally, an individual carrying specifically favorable alleles may be at an advantage related to survival and reproduction (fitness), mediated through natural or sexual selection. Thus differential selection pressure among populations can lead to fast changes and large differences in allele frequency between populations. On the relatively short evolutionary timescale often associated with population processes, migration, drift, and selection are the most important processes. The relative impact of the different evolutionary drivers ultimately determines the genetic composition of populations and the genetic differentiation among them. Thus, migration tends to homogenize allele frequencies among populations, while random genetic drift and differential selection acts to differentiate them. As a rule of thumb, small, isolated populations subject to special environmental conditions tend to show the largest genetic differentiation and are therefore most easily distinguished using genetic tools. Genetic differentiation due to population structure is traditionally measured using a fixation index, "F_{ST}" (Wright, 1950; Weir and Cockerham, 1984). The index ranges from zero to one, where zero denotes no differentiation and one represents fixation of different alleles among populations. As a measure of scale, F_{ST}s among humans on different continents ranges between 0.1 and 0.15 (Jorde and Wooding, 2004).

Types of Population Structure

A prerequisite in order to be able to use genetics to determine the population of origin of seafood is that the marine organism in question display some sort of genetic structuring of populations. In general, there are many evolutionary and ecological models for population structuring of marine organisms, which are beyond the scope of this chapter. However, three crude categories of significant importance for the identification of origin can be recognized (Laikre et al., 2005; Fig. 2.1). First, there is **no genetic differentiation** (panmixia) across the geographical regions of interest, that is, that migration and associated gene flow is sufficient to homogenize populations. This means that genetic tools cannot be used for origin identification as the different regions of the species distribution display non-distinguishable genetic compositions. This may, naturally, be an inherent characteristic of the species in question; however, it may also be an artifact of the sampling strategy and/or the genetic and analytical tools applied (for more details see the following sections). Another type is **continuous genetic change**, that is, allele frequencies shift gradually along a geographical or environmental transect. Accordingly, the genetic compositions at each end of the species distribution are highly genetically differentiated, while intermediate locations display minute and gradual genetic changes. This kind of population structure imposes some problems in relation to determination of origin, as the statistical power associated with referring individuals to specific sites, as opposed to adjacent locations, is expected to be relatively weak. In addition, a significant sampling and genetic typing effort has to be undertaken in order to be able to describe the genetic shape of this "isolation by distance," that is, to establish whether the continuous change is homogenous across the whole distributional area. The final major type is **distinct populations**, where migration among populations is sufficiently small to allow the buildup of distinct genetic differences. This type of population structure not only represents the ideal setting for population-based management and conservation, it also represents the optimal structure for population/origin assignment. As all populations in this scenario are geographically defined and genetically distinct, the population of origin of individuals can be inferred with high probability, dependent on the levels of genetic differentiation among populations. However, as the genetic population represents the reproductive unit, different populations may have distributional areas that significantly differ and overlap outside spawning time. In the latter case, the genetically determined population of origin of an individual fish may provide little information on the geographical origin of the sample. Still the mixture composition, using information from a sufficiently large sample of individuals in concert, may be able to provide insights into the geographical origin. This issue is treated in more detail in a subsequent section.

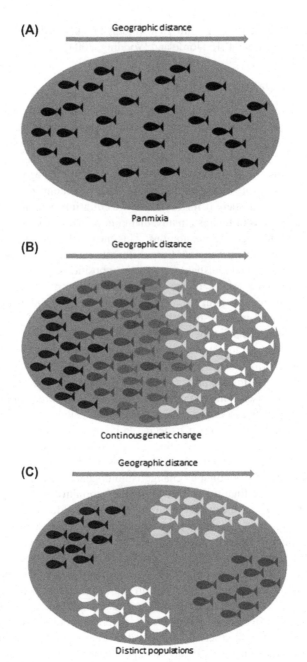

FIGURE 2.1 Three types of population structure for marine organisms (A) no genetic differentiation (B) isolation (C) distinct populations (see text for explanation).

Population Structure of Marine Organisms

The population structure and level of genetic differentiation for important commercial species is of paramount importance for successful origin determination, and subsequently for improved stock management. The level of genetic differentiation among populations (F_{ST}) is typically much lower for marine organisms than for freshwater and anadromous species (Fig. 2.2A, redrawn from Ward et al., 1994). Likewise, it is evident (Fig. 2.2B) that the vast majority of marine fish species display very low levels of genetic differentiation among populations ($F_{ST}<0.03$). The reason for the relatively low levels of genetic differentiation for

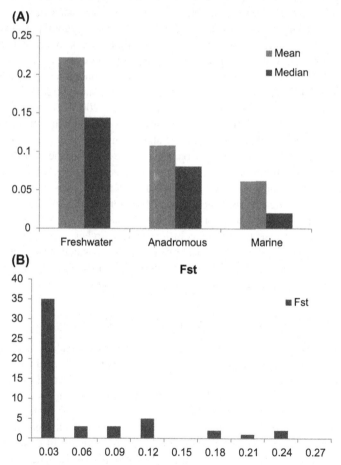

FIGURE 2.2   Levels of genetic differentiation among populations of marine fish. (A) Comparison of genetic differentiation (F_{ST}) among freshwater, anadromous and marine fish. (B) Distribution of F_{ST} values in marine fish. *Redrawn from Ward, R.D., Woodwark, M., Skibinski, D.O.F., 1994. A comparison of genetic diversity levels in marine, freshwater and anadromous fishes. Journal of Fish Biology 44, 213–232.*

"classical marine organisms," including many of our most important commercial species such as clupeoids, gadoids, and scombrids (Nielsen and Kenchington, 2001) relates to a number of inherent characteristics of these species. First, the number of obvious physical barriers in the sea is not so pronounced as in freshwater. For example, while highly mobile marine fish can freely migrate vast distances in the oceans, fish living in a lake are restricted to this particular water body. Likewise, many marine organisms have pelagic eggs and larvae, which can be spread over vast areas by ocean currents before settling. Finally, most marine species have comparatively large (effective) population sizes (Hare et al., 2011) resulting in minute levels of random genetic drift and related low levels of genetic differentiation. However, although it may seem that the oceans are devoid of any physical boundaries, this is not the case. The major oceans are separated by large landmasses, restricting gene flow on a large geographical scale. In addition, factors, such as bathymetry and ocean currents, may serve as barriers to active migration of adult specimens, or act to retain eggs and larvae, so that the juveniles will settle in proximity to the parental population (e.g., see Sinclair and Power, 2015). It has been identified that environmental differences may also restrict migration among populations (Limborg et al., 2009). Thus differences in temperature, salinity, and other environmental factors may define the boundaries between populations. Habitat preference and life history may also restrict gene flow geographically. This phenomenon also sets the scene for the identification of genes subject to differential selection in populations inhabiting different environments. This selection can create vast differences in allele frequencies even in the face of relatively high levels of gene flow, which renders the application of these genes particularly interesting for origin determination. This will be treated in detail in subsequent sections. In conclusion, marine organisms display relatively low levels of genetic differentiation among populations, which in association with the lack of obvious physical boundaries among populations poses a number of challenges for origin determination of seafood products.

PRINCIPLES OF POPULATION ASSIGNMENT

As previously stated, the origin of an individual is equivalent to its genetic population. Thus first, the population of origin has to be determined, and subsequently, this has to be matched with the known or suspected geographical distribution of the population. In order to make this comparison, background knowledge of the populations, their distribution, and their biology must be empirically derived. This may require biological study of an organism and its life history, and/or genotyping of hundreds to thousands of different individuals across a suspected range to identify genetic populations. Where this information is available, population-level assignments can be made. Most often the population of origin of individuals is determined using a method called "individual assignment" (Paetkau et al., 1995; Rannala and Mountain, 1997). In contrast to genetically based species designation (e.g., DNA barcoding), which is categorical with fixed differences in DNA sequences among species, individual assignment (IA) is probabilistic, exploiting differences in allele frequencies among populations. In essence, the method calculates the probability of observing a given multilocus genotype based on the different allele frequencies in a set of reference populations. It is rarely the case that a single genetic marker is sufficient to provide high statistical power for unambiguous assignment of individuals to population. Instead the method relies on combining allele frequency information from a number of genetic markers, thereby increasing the statistical power for inferring population of origin. If significant genetic differences are found among populations, then in theory, any level of statistical certainty should be attainable, by applying more markers. However, in practice, this is limited by the genotyping error rate, cost, and time constraints.

Population assignment is composed of a number of predefined steps allowing rigorous assessment of the most likely population of origin of individuals and the statistical certainty associated with it. First, a set of baseline genetic data has to be retrieved from potential populations of origin (see Fig. 2.3). Typically,

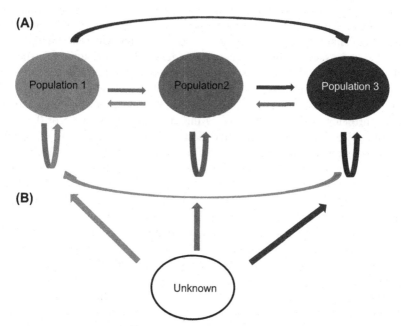

FIGURE 2.3 Principle of individual assignment. (A) "Self-assignment". Likelihoods of observing multilocus genotypes are calculated for all individuals within baseline samples and assigned back to the sample (population) where they have the highest likelihood of occurring. (B) "Assignment of unknown individuals". The likelihood of observing the genotype of an individual of unknown origin is calculated for each of the baseline samples and the individual is assigned to the sample (population) where it has the highest likelihood of occurring (see text for further explanation). *Redrawn from Hansen, M.M., Kenchington, E., Nielsen, E.E., 2001. Assigning individual fish to populations using microsatellite DNA markers: methods and applications. Fish and Fisheries 2, 93–112.*

the markers applied have been single nucleotide polymorphisms (SNPs) or microsatellites (for more information on markers see next section). The second step is to do "self-assignment" to evaluate the statistical power of assignment. For all baseline individuals, the likelihoods of observing their multilocus (across all markers) genotypes in each of the baseline populations are calculated. To avoid biasing allele frequencies, the baseline individual being assigned is commonly excluded in the calculation of allele frequencies in a procedure called "leave one out" (Efron, 1983). The individual is then assigned to the population sample where its genotype has the highest likelihood of occurring based on sample allele frequencies. If the populations are sufficiently genetically distinct, we should expect that most or all baseline individuals are assigned back to their known population of origin. However, the IA procedure entails that all individuals are assigned to a baseline sample no matter how small their likelihoods are. For example, if the population of origin is not included among baseline samples, we may (erroneously) assign the individual to another population included in the baseline. Likewise, individuals may have similar likelihoods in two or more populations, rendering it difficult to state the population of origin with high certainty. So how do we assign a statistical probability to the assignment test? To evaluate the problem of missing baselines Cornuet et al. (1999) devised a method of simulating a large number (>1000) of individuals from each population based on sample allele frequencies to generate a distribution of expected likelihoods for true individuals within the population. The likelihood of each sampled individual is then compared to the simulated distribution of likelihoods and the individual is accepted/rejected from the population if its likelihood is above or below a certain threshold of the distribution (e.g., 0.05 or 0.01). To statistically evaluate, the relative likelihoods of potential alternative origins for a given genotype a number of options are available (Piry et al., 2004). One method is to estimate the relative likelihood scores by dividing the estimated likelihood of observing the genotype in each population by the total likelihood for all populations. Again here a threshold value (e.g., 90% or 95%) can be applied to designate a level above which the assigned population is accepted/rejected as the true population of origin. Alternatively, likelihood ratios between pairs of populations can be calculated, which is often the preferred option for forensic purposes (Ogden, 2008; Ogden and Linacre, 2015). The ideal situation is when the distributions of likelihood ratios from different populations do not overlap and are clearly different from zero (see Fig. 2.4), that is, that the genetic differentiation between populations and the number of markers is sufficient to allow unambiguous assignment. After the assignment power has been assessed, individuals of truly unknown origin can be statistically assigned to the most likely source population among baselines. Again the likelihood of observing each of the "unknown" individual genotypes is calculated for all populations and the individual is assigned to the population where it has the highest likelihood of occurring. Likewise, the statistical evaluation of whether the individual actually could belong to the population where it is assigned and alternative populations

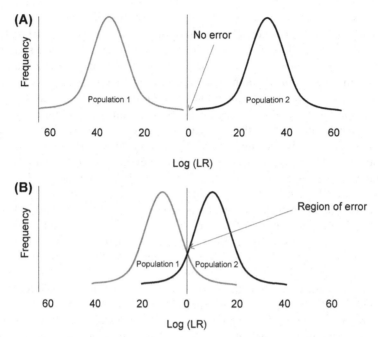

FIGURE 2.4 Distribution of log-likelihood ratios (log LR) for individuals from two different populations where (A) populations are well differentiated and the number of markers are sufficiently high or (B) where genetic differentiation is low and or the number of markers is insufficient for allowing unambiguous assignment. *Redrawn from Ogden, R., Linacre, A., 2015. Wildlife forensic science: a review of genetic geographic origin assignment. Forensic Science International-Genetics 18, 152–159.*

of origin is performed as for the self-assignment of baseline samples. Baseline reference data should be formed from large sample sizes to have appropriate power to discriminate populations with high levels of certainty.

POPULATION OR POINT-OF-ORIGIN IDENTIFICATION IN PRACTICE

Before applying IA methods, there are a number of issues to consider for practical implementation. Of particular concern is the speed and reproducibility of genotyping as well as obtaining sufficiently high statistical power for inferring origin.

Genetic Markers

Mitochondrial DNA is widely used for species identification but only rarely used for origin assignment unless the potential populations of origin are genetically very distinct. MtDNA is (generally) maternally inherited without recombination, so the whole genome is linked and acts effectively as a single genetic marker. This is not optimal for IA (but see Marko et al., 2011 for an example), where the high statistical power of origin assignment relies on the combination of information across multiple genetic markers. Accordingly, IA typically applies a number of nuclear genetic markers, with microsatellites and single-nucleotide polymorphisms (SNPs) as the preferred options. Since the development of assignment tests 20 years ago, most IA studies have relied on the application of microsatellites. Microsatellites consist of tandemly repeated DNA sequence motifs (2–5 base pairs), commonly found in non-coding regions of the genome and often have a high number of alleles per locus (5–50) (Putman and Carbone, 2014). The high number of alleles provides high information content per locus (see section on assignment power below), making them well suited for studies where a relatively modest number of markers (5–15) can be genotyped. However, there are drawbacks of using microsatellites. As their genotyping typically relies on the relative electrophoretic migration of PCR fragments (alleles) of different sizes, they are prone to genotyping errors and are notoriously difficult to calibrate across laboratories (Ellis et al., 2011). In contrast, SNPs are biallelic markers found in all organisms in both coding and noncoding parts of genomes. Many high-throughput methods are available for SNP genotyping, so a relatively higher number of markers can be genotyped for SNPs compared to microsatellites, compensating for the smaller number of segregating alleles. Another major advantage is that SNP genotyping is platform independent, that is, the genotypes identified in different laboratories can be readily compared. Accordingly, SNPs are currently gaining much wider application for IA. A final note of relevance for all genetic markers is that preferably markers imbedded in short DNA fragments should be applied when using low-quality templates, for example from seafood samples that have been degraded due to processing. As a general rule of thumb segments spanning, more than 200 bps have proven difficult to PCR amplify in low-quality historical samples (e.g., see Nielsen and Hansen, 2008). Thus this lends further support to SNPs as the markers of choice as they allow design of marker segments of smaller size.

Assignment Power

The power of assignment tests is determined by a number of factors: (1) the level of genetic differentiation among the sampled populations, (2) the number and polymorphism of genetic markers applied, and (3) the number of sampled populations and individuals (Cornuet et al., 1999; Hansen et al., 2001; Manel et al., 2005). The level of genetic differentiation among populations is obviously very important for assignment power. Simulation studies (Cornuet et al., 1999; Manel et al., 2002) have shown that with an F_{ST} of 0.1 almost 100% correct assignment can be achieved with a relatively limited number of genetic markers

(30 individuals, 10 loci). In contrast, when F_{ST} is low (i.e., 0.01), as for marine fish populations, many more loci have to be used to achieve high assignment power. In the early 2010s, a simulation program "SPOTG" has been developed (Hoban et al., 2013) for choosing the appropriate number of loci and individuals to achieve a desired level of assignment power. By using user generated input on allele frequency distributions and levels of genetic differentiation, the program reports mean and standard deviation on mis-assignment, incorrect inclusion and incorrect exclusion. This approach is highly recommended to design studies using assignment tests.

High Grading of Genetic Markers

IA has traditionally relied on using a limited number of genetic markers only influenced by "neutral" evolutionary forces (random genetic drift and gene flow). However, the advances in genetic sequencing via "next-generation sequencing" is rapidly changing the number of genetic markers available for IA in a given species (Helyar et al., 2011). It is now possible to high-grade assignment panels by choosing the specific genetic markers that show the largest divergence among populations (Bromaghin, 2008) and thus create "minimum marker panels with maximum power" for IA (Nielsen et al., 2012a and below for examples for marine fish). Many of the high divergence markers, will be gene loci directly, or indirectly influenced by selection (i.e., situated in genomic regions, where the different alleles have an influence on the fitness of the individual or "hitchhiking" through linkage with variants under selection). The high level of differentiation is expected to occur due to the process of differential selection imposed by differences in the environment experienced by different populations (Nielsen et al., 2009). Thus selection increases allele frequency differences among populations at these marker loci compared to neutral loci, and as a consequence, they will provide higher power for IA. This means that fewer markers are needed to obtain similar precision, thereby reducing time and costs associated with IA. However, there are also some potential pitfalls associated with high grading (see Anderson, 2010). Differences in population divergence among markers in the baseline (training) samples may simply be caused by the process of sampling baseline individuals. Thus, some allele frequencies may show large differences among baseline samples just by chance and if those markers are deliberately cherry-picked assignment power can be seriously overestimated. Consequently, another assignment procedure is required to evaluate assignment precision. First, each of the baseline samples is split in two (Fig. 2.5), where half of the individuals are used as a baseline or "training" samples and the other half is treated as if the origin was unknown, termed "holdout" samples (Anderson, 2010). The rationale for using these samples is to provide an unbiased estimate of assignment power. In other words to use a sample of individuals of known origin, which has not been used to estimate

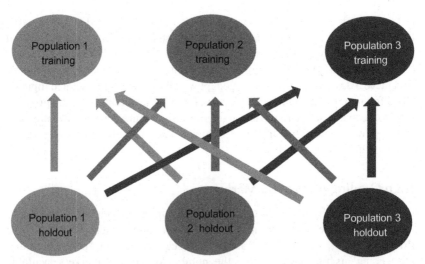

FIGURE 2.5 Procedure for evaluating statistical power when using high grading of loci for IA. Baseline samples are split into "training" and "holdout" samples, where the training sample is used as a baseline for defining population allele frequencies, while the holdout sample is used for evaluation of the statistical power of assignment (see text for more explanation).

allele frequencies within populations, thus resembling a scenario of assigning individuals of truly unknown origin back to baseline (training) samples.

Software for Origin Assignment

A number of statistical analysis methods are available for IA and have been implemented in different software programs. The original frequency-based method developed by Paetkau et al. (1995) has generally been replaced by partly or full probabilistic Bayesian methods (Pritchard, 2000; Piry et al., 2004). The two most commonly applied tools are "GeneClass" (Piry et al., 2004) and "STRUCTURE" (Pritchard, 2000). GeneClass uses a Bayesian approach to estimate baseline allele frequencies, by assuming equal prior probabilities of occurrence of alleles at each locus in each population. This is done to account for potentially missing rare alleles within populations, which has not been detected in individuals sampled for the baseline. The likelihoods of observing a given set of multilocus genotypes in each of the baseline populations and accepting/rejecting that genotype from each of the populations is calculated according to the description in the section on "Principles of population assignment" earlier. The pure Bayesian method implemented in STRUCTURE (Pritchard, 2000), builds on a completely different principle. This method clusters individuals to minimize Hardy–Weinberg and linkage disequilibria within clusters. The rationale behind the model is that it assumes that there is random mating within populations, and therefore, all loci are expected to be in HW and linkage equilibrium. Individuals are then assigned probabilistically back to a population. Multiple populations may be assigned if their genotypes suggest that they are admixed (i.e., represent hybridization between individuals from different populations). In general, the latter method does not perform particularly well for species with low levels of population structuring (Kalinowski, 2011), thus for the assignment of classical marine organisms on an intermediate to local scale STRUCTURE has limited application. However, a program using discriminant analysis of principal components "DAPC" has been developed (Jombart et al., 2010) to identify the best supported number of groupings assumed to represent populations. The method does not rely on a specific genetic model but generates synthetic variables (principal components), using linear combinations of the original variables (alleles) and seeks the variables that maximize differences between groups (discriminant functions). Based on these, DAPC provides membership probabilities of each individual for the different groups. As for any principal component analysis (PCA), retaining too many principal components may lead to overfitting of the data. However, the program includes a procedure for avoiding this. The three described programs represent the most commonly applied, but fundamentally different, approaches for genetic origin assignment. When performing practical IA, it is advisable to test at least a couple of methods to assess the robustness of the assignment. In particular, for marine organisms with shallow population structure, the different assumptions of the approaches may influence the outcome, in particular for cases when the statistical power of IA is low (e.g., few genetic markers, small baseline samples).

CASE STUDIES

Fishing Competition for Atlantic Salmon in Finland

A classical and one of the first examples of the use of origin identification for "seafood" relates to a fishing competition in the Finnish Lake Saimaa in June 1999 (Primmer et al., 2000). In a local fishing competition, one of the participants presented a 5.5 kg salmon to the judges. This salmon was unexpectedly large compared to normal Lake Saimaa salmon. The judges suspected that the salmon could have been purchased or caught elsewhere and submitted tissue samples for genetic analysis. Based on baseline microsatellite genetic data (7 loci) from Lake Saimaa salmon, the authors used the simulation approach (10,000 individuals) in GeneClass to generate an expected distribution of likelihoods for that population (see section on principles of population assignment). They found that the probability of the large salmon belonging to the Lake Saimaa population was very low ($p < .0001$). In contrast, the likelihood of originating from one of the regions in Finland that supply most fish markets was more than 600 times higher. When confronted with the evidence the angler confessed that he had purchased the fish at a local fish shop and criminal charges were laid. The case demonstrates that even if it is not possible to include all potential populations of origin for the baseline data, the exclusion approach can still provide statistically robust tests of specific hypothesis about the point of origin of individual fish.

North Sea or Baltic Sea Cod Sold in Sweden

In 2003, journalists from the Swedish TV4 decided to investigate whether Cod illegally caught in the Baltic Sea were sold under false "North Sea" labels of geographical origin by Swedish fishmongers. The journalists had estimated that fish retailers overall could make more than €500.000 extra annually through deliberate mislabeling, illustrating the large potential for illegal economic revenue. In order to investigate the case, the journalists visited a number of fish shops in Sweden and bought two cod filets labeled as North Sea in origin in each shop. If possible they revisited the shops, resulting 42 samples in total. The claimed origin was documented through oral confirmation by the shop assistants on candid camera. The collected cod samples were subsequently tested against baseline data from North Sea and Baltic Sea Cod (Nielsen et al., 2012b) using 10 microsatellite loci. Genetic differentiation between Baltic and North Sea Cod is relatively high with an F_{ST} of 0.045 (Nielsen et al., 2003). Accordingly, assignment power was high with more than 90% of the Baltic Sea baseline samples correctly assigned. Of the 42 Cod filet samples, 20 were assigned to the North Sea, 17 to the Baltic Sea and the remaining five samples were assigned (based on simulations) neither to the North Sea nor Baltic populations though all 42 had been labeled as North Sea Cod. More than half of the filets were apparently mislabeled, and most of these were likely illegally caught Baltic cod. Interestingly, the pattern of mislabeling was not random. While some shops visited in this study always sold correctly labeled products others always sold mislabeled cod. When confronted with the evidence all shop managers admitted that mislabeling was indeed possible, but claimed that the mislabeling occurred from the wholesalers. When the journalists contacted these companies, none of them agreed to give an interview. When published this investigation created an uproar among consumers in Sweden and neighboring countries as the fish wholesale companies allegedly supplying the mislabeled fish were international.

Using SNPs Under Selection for Origin Identification in Classical Marine Fish

Within the European Union, all traded seafood products require catch certificates stating the origin of catch. Likewise, many products are "eco-labeled" from organizations such as the MSC (Marine Stewardship Council), stating that the fishery from where the products originate is sustainable. Accordingly, there is a need for independent methods to identify the population of origin for commercial fish. However, as genetic differentiation in classical marine fish is commonly relatively low, it can be rather difficult to attain sufficient statistical power for unambiguous origin determination through individual assignment.

The EU supported project FishPopTrace set out to develop high-power assignment tools for four important commercial fish species in European waters: Atlantic Cod, Atlantic Herring, European Sole, and European Hake (Nielsen et al., 2012a). They first used next-generation sequencing to identify SNPs distributed across the genomes for all four species. Subsequently, 1536 (Cod), 281 (Herring), 427 (Sole), and 395 (Hake) SNPs were genotyped in individuals across the species' distributions. SNP's with particularly high F_{ST} were identified as being subject to differential selection and used for designing minimum panels with maximum power for IA, by the process known as "high grading" (see previous section). This was done to allow for rapid processing of a high number of samples within a forensic context. Four species specific scenarios were investigated. For cod, there is a need for methods that can discriminate between North Sea, Barents Sea, and Baltic Sea populations, as their health status varies considerably. Using only eight SNP's with the highest levels of genetic differentiation among populations (F_{ST} between 0.07 and 0.51) correctly assigned all fish back to population of origin. Overall 95% of the individuals had likelihoods that were 1500 times higher for the correct population of origin, thus providing robust results suitable for use in legal proceedings. For herring, no method was found that could distinguish North Sea from Northeast Atlantic Herring, which is important to MSC for certifying fisheries. The 32 highest ranking SNPs (F_{ST} between 0.01 and 0.19) could correctly assign origin for 100% of the Northeast Atlantic and 98% of the North Sea Herring. The true population of origin was always more than three times as likely (maximum seven million times more likely), while the median value was 16,800 times more likely. For Sole, landings in Belgian ports are claimed to originate from the Irish Sea/Celtic Sea. However, they may in fact be caught close to the Belgian coast, which is closer to the market, but where fishing is prohibited to allow rebuilding of the local population. An assay of 50 SNPs with the highest F_{ST} values (between 0.005 and 0.054) correctly assigned 93% to area of origin. On average, individuals were more than 60 times more likely in the population of origin, demonstrating the high power of the method even across a very restricted geographical scale. Finally, for Hake, fishing regulations differ between the Mediterranean and the Atlantic with different legal sizes allowed in the two regions. Thus undersized Atlantic Hake are often misreported as being of Mediterranean origin. Thirteen high F_{ST} SNPs (F_{ST} between 0.08 and 0.29) provided 99% correct assignment to basin of origin. Evaluation of the likelihood of alternative hypotheses of origin showed that 95% of all sampled hake were over 500 times more likely to originate from their basin of sampling than to other basins. Overall, this case demonstrates that the combination of next generation sequencing, SNP development and the application of high grading of markers under differential selection is a very powerful method for developing high-powered IA assays (see also Helyar et al., 2011). However, this represents the results of a large-scale research program that requires substantial financial resources to undertake. Continued development of these assays requires investment in vessel time, research, and analytical costs for determining baseline reference populations for comparisons.

BIOLOGICAL LIMITATIONS AND POTENTIAL PITFALLS OF POINT-OF-ORIGIN IDENTIFICATION

Genetically based individual assignment supports determination of geographical or population of origin of seafood products. The field is developing rapidly in terms of both the type and the number of the genetic markers applied and also with respect to the statistical methods available. In concert, these developments allow origin determination with increasing geographical resolution and precision. However, as mentioned in the introduction to this chapter, there are still a number of factors that could limit the application of point-of-origin identification. These methods are most often applied in natural populations, which implies a need to rely on the far from perfect knowledge of all biological characteristics of the species in question. Lack of population differentiation across the full or main species distribution areas is a major obstacle for genetic origin assignment. This can be an inherent feature of the population, but may also be caused by the choice and number of markers used to attempt description of population differentiation. In addition, many marine species undertake extensive spawning and feeding migrations (e.g., Ruzzante et al., 2006), which may result in extensive mixing of different genetic populations. If these migrations are not documented, erroneous origin determination could take place. Therefore, careful measures must be taken in the design of population-level identification assays to either possess in depth biological knowledge of the species in question, or to be knowledgeable on the potential limitations of applying the method for seafood identification.

In the case of population mixtures, the genetic population signature of individuals may not reveal the geographical origin, while a larger sample consisting of many individuals could provide an overarching sample signature (i.e., proportions of fish from different populations contributing to the mixture), which may expose the origin. This type of analysis is typically conducted using an alternative, but related, approach, a so-called "mixed stock analysis" (Shaklee, 1990), which is optimal for estimating mixture proportions of individuals originating from different populations, rather than the most likely origin of single individuals. However, in order to use such analysis to infer origin, a database of spatiotemporal population mixture signatures for the species in question is required. This may be feasible for specific species and areas, but in general such data are missing. As a conclusion, one should always be cautious when interpreting data on genetic origin assignment. After all, there are few boundaries in the sea and marine organism can swim or potentially get distributed over vast geographical areas. Therefore, it is important to test specific hypothesis about the origin of an individual, in particular for forensic purposes (i.e., prosecutor and defense claims), to see which hypothesis is most strongly supported by the origin assignment and related statistical inferences.

FUTURE PERSPECTIVES FOR POPULATION AND POINT-OF-ORIGIN DETERMINATION

All fields of genetic research are now benefitting from the massive amounts of genomic information generated through "next-generation sequencing" (NGS) technology. This also holds for many seafood species, where full or partial genomes are available for species, such as Atlantic Salmon, Atlantic Cod, and Turbot, (see http://www.ncbi.nlm.nih.gov/genome for a list). Many more are likely to become available within the near future, thus strongly facilitating the development of markers applicable for individual assignment. Sequencing technology is developing extremely quickly, and so in a few years' time, it may be cheaper and faster to apply NGS methods directly for generating genetic data for origin determination. However, at the moment, it is still more practical to develop specific panels of markers with high power for IA for target species and scenarios. Another field where a lot of progress is anticipated is related to the development of portable real-time devices, which at the moment allow for field-based DNA testing in less than 15 min (for an example see the Genie II http://www.optigene.co.uk/instruments/instrument-genie-ii/). The capacity, in terms of numbers of reactions, of these instruments is relatively low, thereby limiting the applicability for origin determination potentially requiring a high number of markers to provide strong statistical inferences for weakly differentiated populations. Still, they may act as a preliminary on-site screening device, where more detailed analysis can subsequently be performed under laboratory conditions. Despite the challenges in applying genetic origin identification to seafood and seafood products outlined here, it is already an applied and generally superior (and sometimes only) method for assigning individuals back to population/geographical origin. It is expected that the genomic revolution will contribute to faster, more cost efficient and precise tools, which can be applied to a wide range of seafood species. This will, however, require that more focus is diverted to provide a better understanding of the biology and genetic population structure for species inhabiting the world's oceans.

REFERENCES

Anderson, E.C., 2010. Assessing the power of informative subsets of loci for population assignment: standard methods are upwardly biased. Molecular Ecology Resources 10, 701–710.

Bromaghin, J.F., 2008. BELS: backward elimination locus selection for studies of mixture composition or individual assignment. Molecular Ecology Resources 8, 568–571.

Børresen, T., 1992. Quality aspects of wild and reared fish. In: Huss, H.H., Jakobsen, M., Liston, J. (Eds.), Quality Insurance in the Fish Industry. Elsevier, Amsterdam, pp. 1–17.

Cadrin, S.X., Kerr, L.A., Mariani, S., 2014. In: Cadrin, S., Kerr, L., Mariani, S. (Eds.), Stock Identification Methods, Stock Identification: An Overview, second ed. Elsevier.

Cornuet, J.M., Piry, S., Luikart, G., Estoup, A., Solignac, M., 1999. Comparison of methods employing multilocus genotypes to select or exclude populations as origins of individuals. Genetics 153, 1989–2000.

Council Regulation (EC) No 1224/2009, 20 November 2009. Official Journal of the European Union L 343 (1), 50.

Council Regulation (EC) No 1005/2008, 29 September 2008. Official Journal of the European Union L 286 (1), 32.

Efron, B., 1983. Estimating the error rate of a prediction rule: improvement on cross-validation. Journal of the American Statistical Association 78, 316–320.

Ellis, J.S., Gilbey, J., Armstrong, A., et al., 2011. Microsatellite standardisation and evaluation of genotyping error in a large multi-partner research programme for conservation of Atlantic salmon (*Salmo salar* L.). Genetica 139, 353–367.

Hansen, M.M., Kenchington, E., Nielsen, E.E., 2001. Assigning individual fish to populations using microsatellite DNA markers: methods and applications. Fish and Fisheries 2, 93–112.

Hare, M.P., Nunney, L., Schwartz, M.K., et al., 2011. Understanding and estimating effective population size for practical application in marine species management. Conservation Biology 25, 438–449.

Helyar, S.J., Hemmer-Hansen, J., Bekkevold, D., et al., 2011. Application of SNPs for population genetics of non-model organism: new opportunities and challenges. Molecular Ecology Resources 11, 123–136.

Higgins, R.M., Danilowicz, B.S., Balbuena, J.A., et al., 2010. Multi-disciplinary fingerprints reveal the harvest location of cod *Gadus morhua* in the northeast Atlantic. Marine Ecology Progress Series 404, 197–206.

Hoban, S., Gaggiotti, O., ConGRESS Consortium, Bertorelle, G., 2013. Sample Planning Optimization Tool for conservation and population Genetics (SPOTG): a software for choosing the appropriate number of markers and samples. Methods in Ecology and Evolution 4, 299–303.

Jombart, T., Devillard, S., Balloux, F., 2010. Discriminant analysis of principal components: a new method for the analysis of genetically structured populations. BMC Genetics 11, 94.

Jorde, L.P., Wooding, S.P., 2004. Genetic variation classification and race. Nature Genetics 36, 28–33.

Kalinowski, S.T., 2011. The computer program STRUCTURE does not reliably identify the main genetic clusters within species: simulations and implications for human population structure. Heredity 106, 625–632.

Laikre, L., Palm, S., Ryman, N., 2005. Genetic population structure of fishes: implications for coastal zone management. Ambio 34, 111–119.

Limborg, M., Pedersen, J.S., Hemmer-Hansen, J., et al., 2009. Genetic population structure of European sprat *Sprattus sprattus*: differentiation across a steep environmental gradient in a small pelagic fish. Marine Ecology Progress Series 379, 213–224.

Manel, S., Berthier, P., Luikart, G., 2002. Detecting wildlife poaching: identifying the origin of individuals with Bayesian assignments tests and multilocus genotypes. Conservation Biology 16, 650–659.

Manel, S., Gaggiotti, O.E., Waples, R.S., 2005. Assignment methods: matching biological questions with appropriate techniques. Trends in Ecology and Evolution 20, 136–142.

Marko, P.B., Nance, H.A., Guynn, K.D., 2011. Genetic detection of mislabeled fish from a certified sustainable fishery. Current Biology 21, 621–622.

Nielsen, E.E., Cariani, A., Mac Aoidh, E., et al., 2012a. Gene-associated markers provide tools for tackling illegal fishing and false eco-certification. Nature Communications 3 Article no 851.

Nielsen, E.E., Hansen, M.M., 2008. Waking the dead: the value of population genetic analyses of historical samples. Fish and Fisheries 9, 450–461.

Nielsen, E.E., Hansen, M.M., Ruzzante, D.E., et al., 2003. Evidence of a hybrid-zone in Atlantic cod (*Gadus morhua*) in the Baltic and the Danish Belt Sea, revealed by individual admixture analysis. Molecular Ecology 12, 1497–1508.

Nielsen, E.E., Hemmer-Hansen, J., Bekkevold, D., 2012b. Development and application of molecular tools to investigate the mislabeling of cod sold in Sweden. In: Hoofar, J. (Ed.), Case Studies in Food Safety and Authenticity: Lessons from Real-life Situations. Woodhead Publishing, Cambridge.

Nielsen, E.E., Hemmer-Hansen, J., Larsen, P.F., Bekkevold, D., 2009. Population genomics of marine fish: identification of adaptive variation in space and time. Molecular Ecology 18, 3128–3150.

Nielsen, E.E., Kenchington, E., 2001. A new approach to prioritizing marine fish and shellfish populations for conservation. Fish and Fisheries 2, 328–343.

Ogden, R., 2008. Fisheries forensics: the use of DNA tools for improving compliance, traceability and enforcement in the fishing industry. Fish and Fisheries 9, 462–472.

Ogden, R., Linacre, A., 2015. Wildlife forensic science: a review of genetic geographic origin assignment. Forensic Science International-Genetics 18, 152–159.

Paetkau, D., Calvert, W., Stirling, I., Strobeck, C., 1995. Microsatellite analysis of population structure in Canadian polar bears. Molecular Ecology 4, 347–354.

Piry, S., Alapetite, A., Cornuet, J.-M., et al., 2004. GENECLASS2: a software for genetic assignment and first-generation migrant detection. The Journal of Heredity 95, 536–539.

Primmer, C.R., Koskinen, M.T., Piironen, J., 2000. The one that did not get away: individual assignment using microsatellite data detects a case of fishing competition fraud. Proceeding of the Royal Society B-Biological Sciences 267, 1699–1704.

Pritchard, J., 2000. Inference of population structure using multilocus genotype data. Genetics 155, 945–959.

Putman, A.I., Carbone, I., 2014. Challenges in analysis and interpretation of microsatellite data for population genetic studies. Ecology and Evolution 4, 4399–4428.

Rannala, B., Mountain, J.L., 1997. Detecting immigration by using multilocus genotypes. Proceedings of the National Academy of Sciences of the United States of America 94, 9197–9201.

Reiss, H., Hoarau, G., Dickey-Collas, M., Wolff, W.J., 2009. Genetic population structure of marine fish: mismatch between biological and fisheries management units. Fish and Fisheries 10, 361–395.

Ruzzante, D.E., Mariani, S., Bekkevold, D., et al., 2006. Biocomplexity in a highly migratory pelagic marine fish, the Atlantic herring. Proceeding of the Royal Society B-Biological Sciences 273, 1459–1464.

Shaklee, J.B., 1990. The electrophoretic analysis of mixed-stock fisheries of Pacific salmon. Progress in Clinical and Biological Research 344, 235–265.

Sinclair, M., Power, M., 2015. The role of "larval retention" in life-cycle closure of Atlantic herring (*Clupea harengus*) populations. Fisheries Research 172, 401–414.

Waples, R.S., Gaggiotti, O., 2006. What is a population? An empirical evaluation of some genetic methods for identifying the number of gene pools and their degree of connectivity. Molecular Ecology 15, 1419–1439.

Ward, R.D., Woodwark, M., Skibinski, D.O.F., 1994. A comparison of genetic diversity levels in marine, freshwater and anadromous fishes. Journal of Fish Biology 44, 213–232.

Weir, B.S., Cockerham, C.C., 1984. Estimating F-statistics for the analysis of population structure. Evolution 36, 1358–1370.

Wright, S., 1950. Genetical structure of populations. Nature 166, 247–249.

Conclusion: DNA-Based Authentication of Shark Products and Implications for Conservation and Management

Robert H. Hanner[1], Amanda M. Naaum[1], Mahmood S. Shivji[2]

[1]University of Guelph, Guelph, ON, Canada; [2]Guy Harvey Research Institute/Save Our Seas Shark Research Centre, Dania Beach, FL, United States

As described in the preceding chapters, a variety of molecular methods are available to address questions concerning the population structure and taxonomic identity of the myriad species that constitute capture fisheries globally. Their application supports seafood authenticity and traceability schemes as well as sustainable fisheries management. Given long generation times and relatively slow reproductive rates, elasmobranchs (sharks and rays) are particularly prone to overexploitation. The unrelenting demand for shark products is unsustainable and many shark fisheries are collapsing. Because of the urgency of addressing this situation, we conclude with an overview of how DNA-based tools are being deployed for the identification of shark products in commercial trade and summarize the relevance of this information for conservation and management. Advances in reference sequence library construction, population-level identification methods, and instrumentation platforms, together with declining costs of conducting molecular diagnostic tests, will enhance the uptake of these tools for seafood authentication and traceability. However, as this chapter demonstrates, they are already improving our ability to monitor patterns of exploitation and yield greater transparency in the industry. The results highlight the urgency of enforcing existing regulations and promoting additional measures to conserve the world's shark fisheries.

THE NEED FOR IMPROVED TESTING

Sharks are harvested through both targeted capture fisheries and landed as bycatch from other commercial fishing practices. A quarter of shark and ray species are now categorized as threatened, endangered, or critically endangered

Seafood Authenticity and Traceability. http://dx.doi.org/10.1016/B978-0-12-801592-6.00009-7

according to the "red list" maintained by the International Union for the Conservation of Nature (IUCN; Dulvy et al., 2014). Sadly, this situation seems to have little impact in reducing the demand for shark products (Clarke et al., 2006a). Due to the lucrative financial rewards, as with many other seafood species, the illegal unregulated and unreported (IUU) catch is substantial (Clarke et al., 2006b; Dulvy et al., 2008). Beyond the economic concerns associated with the collapse of this industry due to overexploitation, there are also major conservation concerns about the impact of removing these apex predators from ocean ecosystems (Myers et al., 2007; Heithaus et al., 2008; Ferretti et al., 2010).

International measures have been suggested to mitigate the impacts of shark fishing. Diverse policies involving quotas on specific species and creation of marine reserves exist (Shiffman and Hammerschlag, 2016), and it is important to note that few of them can be enforced without access to accurate catch statistics. Species identification, and wherever possible, population-level data should be collected on all landed and traded shark catches, yet this is very difficult without access to appropriate expertise and analytical tools. Although training tools exist to help regulators identify fins using morphology (e.g., sharkfinid.com), their application can be problematic because the training tools are generally focused on those species already identified to be at risk. While identifying whole sharks can be easier, this still requires extensive training, and similar species can be easily confused for one another, particularly when dealing with juveniles. This generally results in landings being identified using common names or even more generally as "large," "pelagic," and other nonspecific descriptions. These, while more informative than forgoing the collection of information completely, nevertheless do not meet the requirements for the catch data needed to promote the sustainable management of shark fisheries. Finally, identification after processing into fillets, dried, or powdered shark products cannot be carried out by visual means.

Modern DNA-based methods are important for species identification and population discrimination, both of which are important to fisheries management. In some cases, both are needed to address the management of a single type of fishery. Many of the methods described earlier in this text have been applied to sharks and collectively they illustrate a range of important techniques and their application. Comprehensive reviews have been previously published (Dudgeon et al., 2012; Rodrigues-Filho et al., 2012). Here we highlight some of the ways DNA-based methods are being applied to improve the management of shark fisheries.

SPECIES IDENTIFICATION OF SHARKS

Because different species differ in their abundances, life histories, geographic ranges, and patterns of migration, they also differ in their susceptibility to exploitation (e.g., Smith et al., 1998; Cortés, 2002). As noted earlier, many policies are designed to protect individual species, and therefore their identification is critical to enforcement. Identification of landed sharks is often superficial with multiple inconsistent names applied (e.g., Velez-Zuazo et al., 2015), which makes management efforts imprecise since stocks cannot be assessed without this information (Clarke et al., 2007). Notably, DNA analysis can be used to make identifications that are otherwise difficult, and has improved catch monitoring data for sharks by establishing baseline identification error rates associated with morphological identifications made by onboard observers (e.g., Tillett et al., 2012). Accurate identification is critical for the management of other policies not directly related to protection of specific species as well. For example, some policies require that shark fins retained on board do not exceed the relevant number of carcasses caught, yet do not stipulate that fins remain attached to the carcass, creating an opportunity for fins of more desirable species to be substituted for those that belonged to carcasses kept (Shiffman and Hammerschlag, 2016). DNA analysis can be used in these cases to determine that the fins match the carcasses.

Sequencing mitochondrial DNA, particularly targeting the DNA barcode region of COI, has aided identification of shark fins (Holmes et al., 2009). Numerous studies have helped provide accurate baseline data for landed species in various fisheries associated with Egypt (Moftah et al., 2011), India (Bineesh et al., 2016), Madagascar (Doukakis et al., 2011; Robinson and Sauer, 2013), Mexico (Espinosa et al., 2015), and Peru (Velez-Zuazo et al., 2015), among others. Barcoding has also revealed the presence of previously unknown species in certain geographies, such as the ragged-tooth shark from the south-west Atlantic (Santander-Neto et al., 2011), the sandbar shark from Indian waters (Sutaria et al., 2015), and the porbeagle from Argentinian waters near Buenos Aires (Mabragaña et al., 2015). This approach has also been used to identify several examples of threatened shark species landed or sold in food markets and instances of species mislabeling of shark products (Shivji et al., 2005; Barbuto et al., 2010; Jabado et al., 2015a,b; Sembiring et al., 2015; Moore et al., 2013; Naaum and Hanner, 2015; Wong et al., 2009; Holmes et al., 2009; Asis et al., 2016; Liu et al., 2013; Lambarri et al., 2015). Other mitochondrial gene regions have also been suggested as targets for shark identification and have been used to differentiate species of sharks, at times in concert with COI (Chapman et al., 2003; Naylor et al., 2012; Ovenden et al., 2010; Mendoca et al., 2011; Henderson et al., 2016). Mitochondrial markers like these do have some drawbacks, in that they do not directly address potential hybridization due to maternal inheritance of mitochondrial DNA. However, when used in combination with nuclear markers, this limitation can be overcome. For example, together they have been used to demonstrate hybridization between Australian and common black tip sharks (Morgan et al., 2012), which was thought to be unlikely among shark species since they exhibit internal fertilization. Multiplex species-specific PCR using portions of a nuclear gene, ITS2, can also be used to identify hybridization, and potential heteroplasmy or pseudogenes, that analysis of mitochondrial DNA alone cannot (e.g., Pinhal et al., 2012). This method is relatively simple, and requires less-sophisticated equipment than sequencing analysis. The equipment is small enough that it could potentially be transported in a mobile lab.

An even simpler and more portable option would be to move to real-time PCR. Although this increases costs compared to end-point PCR, the potential for simplification of the protocols means that minimal training of personnel is required and the elimination of post-PCR steps greatly reduces the chance of laboratory contamination and time required for analysis. Real-time PCR has been applied to identification of shark species (e.g., Wang et al., 2015; Morgan et al., 2011). With both end-point PCR and real-time PCR for species identification, species-specific primers for every target species must be individually designed and validated. These methods provide no information about the identity of a sample other than the target species, in which case the sample will require testing with additional species-specific primers or DNA sequencing analysis for a positive identification. However, they do provide a simple and cost-effective means to identify known target species.

POPULATION IDENTIFICATION OF SHARKS

Even more detailed than species identification, and required for adequate stock management, population-level identification using DNA has also been demonstrated for sharks. In many cases, population assessments show much greater population structuring than previously anticipated (e.g., Karl et al., 2011; Ahonen et al., 2009; Sodre et al., 2012). There are also examples where little to no population discrimination is evident (e.g., Ovenden et al., 2011; Veríssimo et al., 2011; Karl et al., 2012) either due to panmixia or because the genetic marker in question evolves too slowly to provide adequate resolution of the underlying population structure, highlighting the importance of using multiple loci to detect it. Management units (e.g., FAO areas) represent putative populations, but the biological information required to validate them as such remains absent for many species. Therefore, continued research on DNA methods for population identification will directly improve the ability to track and manage shark fisheries.

APPLICATION TO FISHERIES MANAGEMENT

Monitoring the fin trade is one avenue to assess information on patterns of species exploitation. Forensic data is crucial to improve monitoring of the shark fin trade, and molecular methods are now being applied by academics, NGOs, and regulatory agencies for this purpose. The utility of DNA data to inform recovery strategies was shown in deep-sea gulper sharks (Daley et al., 2012). DNA identification can aid in rapid identification of landed specimens to enforce regulations, while knowledge of population structure is helping to provide more accurate information for stock management and may help build a more sustainable industry. Notably, multiple stakeholders came together for the first time to document the identity of shark fins being exported from the Mexican Pacific using barcoding (Espinosa et al., 2015). Similarly, the ITS marker was used to demonstrate the presence of at least 10 pelagic shark species in the north-central Chilean fin trade, greatly expanding the list of just four species included in official government landing records (Sebastian et al., 2008). For highly processed fin products where DNA may be degraded, a mini-barcoding assay can be applied to detect shark species that are listed on the Appendices of the Convention on International Trade in Endangered Species (CITES; Fields et al., 2015).

Ongoing research is still needed, particularly for population-level identification, but also for continued development of sequence libraries for elasmobranchs (Becker et al., 2011), which is important for both barcoding and for developing other types of testing methods using species-specific primers. Information from developing nations is also critical. Regions such as Indonesia and Taiwan that most heavily harvest sharks are not well represented in the scientific literature related to shark biology, management, and identification (Momigliano and Harcourt, 2014). All of the methods mentioned for the identification of sharks, and other seafood, rely on access to data, including reference sequences that cover the geographic range of each species and have sufficient sampling to accurately assess inter- versus intraspecific variation. Efforts to continue building the species occurrence, DNA sequence databases and research baselines of both biology and sequence data are critical for other seafood species as well. Shellfish and crustaceans, for example, are underrepresented in the DNA barcode databases compared to fishes, suggesting they are good targets for further study if we want to accurately identify all seafood species and not just bony fishes and sharks.

Emerging methods are likely to increase the capability for scientific testing, but amendments to existing policies and regulations in support of these methods are needed. Access to better testing, and policy enforcement for developing nations is also required. Stakeholders from diverse backgrounds in the seafood industry, would benefit from working together to ensure these emerging markets have access to, and are using these tools, as their participation is critical for the sustainable management of seafood stocks generally and sharks specifically. A focus on methodological standardization, data sharing, and development of harmonized international guidelines for seafood identification will help. Moreover, support for the development of specimen collections and related repositories of genetic reference material (e.g., Ocean Genome Legacy) are necessary for the development, validation, and proficiency testing requirements of new regulatory standards concerning DNA-based identification of populations and species. International collaboration and dedicated funding are required to achieve this vision of a comprehensive framework for managing the last remaining wild capture fisheries on Earth. The combination of molecular methods and modern digital information systems promise a better future for seafood authentication and traceability (Costa and Carvalho, 2007). This book is dedicated to the hardworking individuals (past, present, and future) who are undertaking this cause.

REFERENCES

Ahonen, H., Harcourt, R.G., Stow, A.J., 2009. Nuclear and mitochondrial DNA reveals isolation of imperilled grey nurse shark populations (*Carcharias taurus*). Molecular Ecology 18, 4409–4421. http://dx.doi.org/10.1111/j.1365-294X.2009.04377.x.

Asis, A.M.J.M., Lacsamana, J.K.M., Santos, M.D., 2016. Illegal trade of regulated and protected aquatic species in the Philippines detected by DNA barcoding. Mitochondrial DNA 27, 659–666.

Barbuto, M., Galimberti, A., Ferri, E., Labra, M., Malandra, R., Galli, P., Casiraghi, M., 2010. DNA barcoding reveals fraudulent substitutions in shark seafood products: the Italian case of "palombo" (*Mustelus* spp.). Food Research International. 43 (1), 376–381. http://doi.org/10.1016/j.foodres.2009.10.009.

Becker, S., Hanner, R., Steinke, D., 2011. Five years of FISH-BOL: brief status report. Mitochondrial DNA 22 (Suppl. 1), 3–9.

Bineesh, K.K., Gopalakrishnan, A., Akhilesh, K.V., Sajeela, K.A., Abdussamad, E.M., Pillai, N.G.K., Ward, R.D., 2016. DNA barcoding reveals species composition of sharks and rays in the Indian commercial fishery. Mitochondrial DNA Part A 1–15.

Chapman, D.D., Abercrombie, D.L., Douady, C.J., Pikitch, E.K., Stanhopen, M.J., Shivji, M.S., 2003. A streamlined, bi-organelle, multiplex PCR approach to species identification: application to global conservation and trade monitoring of the great white shark, *Carcharodon carcharias*. Conservation Genetics 4, 415–425.

Clarke, S.C., Magnussen, J.E., Abercrombie, D.L., McAllister, M., Shivji, M.S., 2006a. Identification of shark species composition and proportion in the Hong Kong shark fin market using molecular genetics and trade records. Conservation Biology 20, 201–211.

Clarke, S.C., McAllister, M.K., Milner-Gulland, E.J., Kirkwood, G.P., Michielsens, C.G.J., Agnew, D.J., Pikitch, E.K., Nakano, H., Shivji, M.S., 2006b. Global estimates of shark catches using trade records from commercial markets. Ecology Letters 9, 1115–1126.

Clarke, S., Milner-Gulland, E.J., Bjørndal, T., 2007. Social, economic and regulatory drivers of the shark fin trade. Marine Resource Economics 22, 305–327.

Cortés, E., 2002. Incorporating uncertainty into demographic modeling: application to shark populations and their conservation. Conservation Biology 16, 1048–1062.

Costa, F.O., Carvalho, G.R., 2007. The barcode of life initiative: synopsis and prospective societal impacts of DNA barcoding of fish. Life Sciences Society and Policy 3 (2), 29.

Daley, R.K., Appleyard, S.A., Koopman, M., 2012. Genetic catch verification to support recovery plans for deepsea gulper sharks (genus *Centrophorus*, family *Centrophoridae*) – an Australian example using the 16S gene. Marine and Freshwater Research. 63, 708–714. http://dx.doi.org/10.1071/MF11264.

Doukakis, P., Hanner, R., Shivji, M., Bartholomew, C., Chapman, D., Wong, E., Amato, G., 2011. Applying genetic techniques to study remote shark fisheries in northeastern Madagascar. Mitochondrial DNA. 22 (Suppl. 1), 15–20. http://doi.org/10.3109/19401736.2010.526112.

Dudgeon, C.L., Blower, D.C., Broderick, D., Giles, J.L., Holmes, B.J., Kashiwagi, T., Krück, N.C., Morgan, J.A.T., Tillett, B.J., Ovenden, J.R., 2012. A review of the application of molecular genetics for fisheries management and conservation of sharks and rays. Journal of Fish Biology 80, 1789–1843. http://dx.doi.org/10.1111/j.1095-8649.2012.03265.x.

Dulvy, N.K., Baum, J., Clarke, S., Compagno, L.J.V., Cortés, E., Domingo, A., Fordham, S., Fowler, S., Francis, M.P., Gibson, C., Martínez, J., Musick, J.A., Soldo, A., Stevens, J.D., Valenti, S., 2008. You can swim but you can't hide: the global status and conservation of oceanic pelagic sharks and rays. Aquatic Conservation: Marine and Freshwater Ecosystems 18, 459–482.

Dulvy, N.K., Fowler, S.L., Musick, J.A., Cavanagh, R.D., Kyne, P.M., Harrison, L.R., Carlson, J.K., Davidson, L.N., Fordham, S.V., Francis, M.P., Pollock, C.M., Simpfendorfer, C.A., Burgess, G.H., Carpenter, K.E., Compagno, L.J., Ebert, D.A., Gibson, C., Heupel, M.R., Livingstone, S.R., Sanciangco, J.C., Stevens, J.D., Valenti, S., White, W.T., 2014. Extinction Risk and Conservation of the World's Sharks and Rays. Elife. http://dx.doi.org/10.7554/eLife.00590.001.

Espinosa, H., Lambarri, C., Martínez, A., Jiménez, A., 2015. A case study of the forensic application of DNA Barcoding to sharkfin identification in the Mexican Pacific. DNA Barcodes 3 (1), 94–97.

Ferretti, F., Worm, B., Britten, G.L., Heithaus, M.R., Lotze, H.K., 2010. Patterns and ecosystem consequences of shark declines in the ocean. Ecology Letters 13, 1055–1071. http://dx.doi.org/10.1111/j.1461-0248.2010.01489.x.

Fields, A.T., Abercrombie, D.L., Eng, R., Feldheim, K., Chapman, D.D., 2015. A novel mini-DNA barcoding assay to identify processed fins from internationally protected shark species. PLoS One. 10 (2), e0114844. http://doi.org/10.1371/journal.pone.0114844.s001.

Heithaus, M.R., Frid, A., Wirsing, A.J., Worm, B., 2008. Predicting ecological consequences of marine top predator declines. Trends in Ecology and Evolution 23 (4), 202–210.

Henderson, A.C., Reeve, A.J., Jabado, R.W., Naylor, G.J.P., 2016. Taxonomic assessment of sharks, rays and guitarfishes (Chondrichthyes:Elasmobranchii) from south-eastern Arabia, using the NADH dehydrogenase subunit 2 (NADH2) gene. Zoological Journal of the Linnean Society 176, 399–442.

Holmes, B.H., Steinke, D., Ward, R.D., 2009. Identification of shark and ray fins using DNA barcoding. Fisheries Research. 95 (2–3), 280–288. http://doi.org/10.1016/j.fishres.2008.09.036.

Jabado, R.W., Al Ghais, S.M., Hamza, W., Shivji, M.S., Henderson, A.C., 2015a. Shark diversity in the Arabian/Persian Gulf higher than previously thought: insights based on species composition of shark landings in the United Arab Emirates. Marine Biodiversity 45, 719–731.

Jabado, R.W., Al Ghais, S.M., Hamza, W., Henderson, A.C., Spaet, J.L.Y., Shivji, M.S., Hanner, R.H., 2015b. The trade in sharks and their products in the United Arab Emirates. Biological Conservation 181, 190–198.

Karl, S.A., Castro, A.L.F., Lopez, J.A., Charvet, P., Burgess, G.H., 2011. Phylogeography and conservation of the bull shark (*Carcharhinus leucas*) inferred from mitochondrial and microsatellite DNA. Conservation Genetics 12, 371–382. http://dx.doi.org/10.1007/s10592-010-0145-1.

Karl, S.A., Castro, A.L., Garla, R.C., 2012. Population genetics of the nurse shark (*Ginglymostoma cirratum*) in the western Atlantic. Marine Biology 159 (3), 489–498. http://dx.doi.org/10.1007/s00227-011-1828-y.

Lambarri, C., Espinosa, H., Martinez, A., Hernandez, A., 2015. Cods for sale. Do we know what we are buying? DNA Barcodes 3, 27–29.

Liu, S.V., Chan, C.C., Lin, O., Hu, C., 2013. DNA barcoding of shark meats identify species composition and CITES-listed species from the markets in Taiwan. PLoS One. http://dx.doi.org/10.1371/journal.pone.0079373.

Mabragaña, E., Astarloa, J.M., Lucifora, L.O., 2015. A record of the Porbeagle, Lamna nasus, in coastal waters of Buenos Aires (Argentina) confirmed by DNA barcoding. DNA Barcodes 3 (1), 139–143.

Mendoca, F.F., Oliveira, C., Burgess, G., Coelho, R., Piercy, A., Gadig, O.B.F., Foresti, F., 2011. Species delimitation in sharpnose sharks (genus *Rhizoprionodon*) in the western Atlantic Ocean using mitochondrial DNA. Conservation Genetics 12, 193–200. http://dx.doi.org/10.1007/s10592-010-0132-6.

Moftah, M., Aziz, S.H.A., Elramah, S., Favereaux, A., 2011. Classification of sharks in the Egyptian Mediterranean waters using morphological and DNA barcoding approaches. PLoS One. 6 (11). http://doi.org/10.1371/journal.pone.0027001.

Momigliano, P., Harcourt, R., 2014. Scientists focusing on the wrong sharks in the wrong places. The Conversation: US Edition. May 14, 2014 http://theconversation.com/scientists-focusing-on-the-wrong-sharks-in-the-wrong-places-26245.

Moore, A.B., Almojil, D., Harris, M., Jabado, R.W., White, W.T., 2013. New biological data on the rare, threatened shark *Carcharhinus leiodon* (Carcharhinidae) from the Persian Gulf and Arabian Sea. Marine and Freshwater Research 65, 327–332 http://dx.doi.org/10.1071/MF13160.

Morgan, J.A.T., Welch, D.J., Harry, A.V., Street, R., Broderick, D., Ovenden, J.R., 2011. A mitochondrial species identification assay for Australian blacktip sharks (*Carcharhinus tilstoni*, *C. limbatus* and *C. amblyrhynchoides*) using real-time PCR and high-resolution melt analysis. Molecular Ecology Resources 11, 813–819.

Morgan, J.A.T., Harry, A.V., Welch, D.J., Street, R., White, J., Geraghty, P.T., Macbeth, W.G., Tobin, A., Simfendorfer, C.A., Ovenden, J.R., 2012. Detection of interspecies hybridisation in Chondrichthyes: hybrids and hybrid offspring between Australian (*Carcharhinus tilstoni*) and common (*C. libatus*) blacktip shark found in an Australian fishery. Conservation Genetics 13, 455–463.

Myers, R.A., Baum, J.K., Shepherd, T.D., Powers, S.P., Peterson, C.H., 2007. Cascading effects of the loss of apex predatory sharks from a coastal ocean. Science 315, 1846–1850.

Naaum, A.M., Hanner, R., 2015. Community engagement in seafood identification using DNA barcoding reveals market substitution in Canadian seafood. DNA Barcodes 3 (1), 74–79.

Naylor, G.J.P., Caira, J.N., Jensen, K., Rosana, K.A.M., White, W.T., Last, P.R., 2012. A DNA sequence–based approach to the identification of shark and ray species and its implications for global elasmobranch diversity and parasitology. Bulletin of the American Museum of Natural History.

Ovenden, J.R., Morgan, J.A.T., Kashiwagi, T., Broderick, D., Salini, J., 2010. Towards better management of Australia's shark fishery: genetic analyses reveal unexpected ratios of cryptic blacktip species *Carcharhinus tilstoni* and *C. limbatus*. Marine and Freshwater Research 61, 253–262.

Ovenden, J.R., Morgan, J.A.T., Street, R., Tobin, A., Simpfendorfer, C., Machbeth, W., Welch, D., 2011. Negligible evidence for regional genetic population structure for two shark species *Rhizoprionodon acutus* (Rüppell, 1837) and *Sphyrna lewini* (Griffith & Smith, 1834) with contrasting biology. Marine Biology 158, 1497–1509.

Pinhal, D., Shivji, M.S., Nachtigall, P.G., Chapman, D.D., Martins, C., 2012. A streamlined DNA tool for global identification of heavily exploited coastal shark species (genus *Rhizoprionodon*). PLoS One. http://dx.doi.org/10.1371/journal.pone.0034797.

Robinson, L., Sauer, W.H.H., 2013. A first description of the artisanal shark fishery in northern Madagascar: implications for management. African Journal of Marine Science 35 (1).

Rodrigues-Filho, L.F., Pinhal, D., Sodré, D., Vallinoto, M., 2012. Shark DNA Forensics: Applications and Impacts on Genetic Diversity. INTECH Open Access Publisher.

Santander-Neto, J., Faria, V.V., Castro, A.L., Burgess, G.H., 2011. New record of the rare raggedtooth shark, *Odontaspis ferox* (Chondrichthyes:Odontaspidae) from the south-west Atlantic identified using DNA bar coding. Marine Biodiversity Records 4, e75.

Sebastian, H., Haye, P.A., Shivji, M.S., 2008. Characterization of the pelagic shark-fin trade in north-central Chile by genetic identification and trader surveys. Journal of Fish Biology. 73 (10), 2293–2304. http://doi.org/10.1111/j.1095-8649.2008.02016.x.

Sembiring, A., Pertiwi, N.P.D., Mahardini, A., Wulandari, R., Kurniasih, E.M., Kuncoro, A.W., Cahyani, N.K.D., Anggoro, A.W., Ulfa, M., Madduppa, H., Carpenter, K.E., Barber, P.H., Mahardika, G.N., 2015. DNA barcoding reveals targeted fisheries for endangered sharks in Indonesia. Fisheries Research 164, 130–134.

Shiffman, D.S., Hammerschlag, N., 2016. Shark conservation and management policy: a review and primer for non-specialists. Animal Conservation. n/a–n/a http://doi.org/10.1111/acv.12265.

Shivji, M.S., Chapman, D.D., Pikitch, E.K., Raymond, P.W., 2005. Genetic profiling reveals illegal international trade in fins of the great white shark, *Carcharodon carcharias*. Conservation Genetics 6, 1035–1039.

Smith, S.E., Au, D.W., Show, C., 1998. Intrinsic rebound potentials of 26 species of Pacific sharks. Marine and Freshwater Research 49, 663–678.

Sodre, D., Rodrigues-Filho, L.F.S., Souza, R.F.C., Rego, P.S., Scheider, H., Sampaio, I., Vallinoto, M., 2012. Inclusion of South American samples reveals new population structuring of the blacktip shark (*Carcharhinus limbatus*) in the western Atlantic. Genetics and Molecular Biology. 35. http://dx.doi.org/10.1590/S1415-47572012005000062.

Sutaria, D., Parikh, A., Barnes, A., Jabado, R.W., 2015. First record of the sandbar shark, *Carcharhinus plumbeus*,(Chondrichthyes:Carcharhiniformes:Carcharhinidae) from Indian waters. Marine Biodiversity Records 8, e126.

Tillett, B.J., Field, I.C., Bradshaw, C.J.A., Johnson, G., Buckworth, R.C., Meekan, M.G., Ovenden, J.R., 2012. Accuracy of species identification by fisheries observers in a north Australian shark fishery. Fisheries Research. 127–128, 109–115. http://doi.org/10.1016/j.fishres.2012.04.007.

Velez-Zuazo, X., Alfaro-Shigueto, J., Mangel, J., Papa, R., Agnarsson, I., January 2015. What barcode sequencing reveals about the shark fishery in Peru. Fisheries Research. 161, 34–41. ISSN: 0165-7836. http://dx.doi.org/10.1016/j.fishres.2014.06.005.

Veríssimo, A., McDowell, J.R., Graves, J.E., 2011. Population structure of a deep-water squaloid shark, the Portuguese dogfish (*Centroscymnus coelolepis*). ICES Journal of Marine Science 68, 555–563.

Wang, D.L., Lie, D.H., Xu, L.Z., Guo, X.D., 2015. A real-time pcr method for the detection of shark-origin components in shark fin products. Modern Food Science and Technology 31, 289–293.

Wong, E.H.-K., Shivji, M.S., Hanner, R.H., 2009. Identifying sharks with DNA barcodes: assessing the utility of a nucleotide diagnostic approach. Molecular Ecology Resources 9 (Suppl. 1), 243–256.

Index

Printed in the United States
By Bookmasters